Six-Minute Solutions
for Civil PE Exam
Construction Problems

Elaine Huang, PE

The Power to Pass®
www.ppi2pass.com

Professional Publications, Inc. • Belmont, California

Benefit by Registering This Book with PPI

- Get book updates and corrections.
- Hear the latest exam news.
- Obtain exclusive exam tips and strategies.
- Receive special discounts.

Register your book at **ppi2pass.com/register**.

Report Errors and View Corrections for This Book

PPI is grateful to every reader who notifies us of a possible error. Your feedback allows us to improve the quality and accuracy of our products. You can report errata and view corrections at **ppi2pass.com/errata**.

SIX-MINUTE SOLUTIONS FOR CIVIL PE EXAM CONSTRUCTION PROBLEMS

Current printing of this edition: 4

Printing History

date	edition number	printing number	update
Sep 2012	1	2	Minor corrections.
May 2014	1	3	Minor corrections.
Dec 2015	1	4	Minor cover updates.

Printed in the United States of America.

PPI
1250 Fifth Avenue
Belmont, CA 94002
(650) 593-9119
ppi2pass.com

ISBN: 978-1-59126-370-8

Library of Congress Control Number: 2011944630

F E D C B A

Table of Contents

About the Author

Elaine Huang, PE, is a registered civil engineer in multiple states. Ms. Huang holds a bachelor of science degree in civil and environmental engineering from the University of Massachusetts Amherst and a master of engineering degree from the Massachusetts Institute of Technology. She has designed commercial and industrial structures, and has managed the construction of concrete, steel, masonry, and wood structures. Most recently, Ms. Huang has worked for the defense industry.

Ms. Huang is a solo concert pianist and has performed in many recitals. She also enjoys playing chamber music repertoires with local musicians. Her other hobbies include traveling, running, and horseback riding. She resides outside of Boston with her beloved puppy, Bailey.

Preface and Acknowledgments

Six-Minute Solutions for Civil PE Exam Construction Problems contains problems inspired by my own experience studying for the Principles and Practice of Engineering examination (PE exam). I prepared for my PE exam using the *Civil Engineering Reference Manual* (CERM, also published by PPI). Often while reviewing CERM, I found myself amazed at both the quality and quantity of civil engineering topics covered. I wondered how such a book comes to be written, and after passing my PE exam, I contemplated writing a book to help other engineers pass their PE exam as CERM had helped me pass mine. So, when PPI contacted me about writing *Six-Minute Solutions for Civil PE Exam Construction Problems*, I jumped at the chance.

The topics covered in *Six-Minute Solutions* coincide with those subject areas identified by the National Council of Examiners for Engineering and Surveying (NCEES) for the construction depth module of the civil PE exam. Included among these problem topics are Earthwork Construction and Layout; Estimating Quantities and Costs; Scheduling; Material Quality Control and Production; Temporary Structures; Worker Health, Safety, and Environment; and Other Topics.

This book's problems are representative of the type and difficulty of those problems you will encounter on the PE exam. The problems are both conceptual and practical, and they are written to provide varying levels of difficulty. Though you probably won't encounter problems on the exam exactly like those presented here, reviewing these problems and solutions will increase your familiarity with the form, content, and solution methods of exam-like problems. This preparation will help you considerably during the exam.

Throughout the writing process, the people at PPI assisted me tremendously. They impressed me with their devotion, enthusiasm, and professionalism throughout this book's development. It was an honor to work with Katie Throckmorton, acquisitions assistant, who contacted me about this opportunity and who first reviewed my manuscript; Sarah Hubbard, director of product development and implementation, who guided me through the publishing process; and Jenny Lindeburg King, editorial project manager, who not only edited my problems and solutions to be concise and user-friendly, but checked their accuracy as well.

Thanks also go to Michael R. Lindeburg, PE, for inspiring me to write this book, as well as to Cathy Schrott, production services manager; Megan Synnestvedt, editorial project manager; Lisa Devoto, copy editor; Tyler Hayes, copy editor; Chelsea Logan, copy editor; Scott Marley, copy editor; Magnolia Molcan, copy editor; Bonnie Thomas, copy editor; Kate Hayes, production associate; Todd Fisher, calculation checker; Tom Bergstrom, illustrator; and Amy Schwertman La Russa, illustrator and cover designer.

In addition to the PPI staff, I would like to acknowledge Thomas W. Schreffler, PE, for his technical review of my manuscript. His critical eye and valuable comments have helped improve this book.

And, I would be remiss if I didn't offer my greatest gratitude to all my undergraduate and graduate professors, whose passion and pride for the civil engineering profession encouraged my interest in engineering. From their teachings, I gained a solid engineering foundation that I continue to build upon in my daily career as an engineer.

Finally, though the problems and solutions in this book have been carefully prepared and reviewed to ensure they are appropriate and understandable, if you do find an error or discover an alternative, more efficient way to solve a problem, I encourage you to let me know. You can report errors, offer suggestions, and keep up with the changes made to this book through PPI's errata website, **www.ppi2pass.com/errata**.

All the best in your endeavors as an engineer.

Elaine Huang, PE

Introduction

EXAM FORMAT

The Principles and Practice of Engineering examination (PE exam) in civil engineering is an eight-hour exam divided into a morning and an afternoon session. The morning session is known as the "breadth" exam and the afternoon is known as the "depth" exam.

The morning session includes 40 problems from all of the five civil engineering subdisciplines (construction, geotechnical, structural, transportation, and water resources and environmental). Each subdiscipline represents about 20% of the problems. As the "breadth" designation implies, morning session problems are general in nature and wide-ranging in scope.

The afternoon session allows the examinee to select a depth exam module from one of the five subdisciplines. The 40 problems included in the afternoon session require more specialized knowledge than do those in the morning session.

All problems from both the morning and afternoon sessions are multiple choice. They include a problem statement with all required defining information, followed by four logical answer options. Only one of the four options is correct. Nearly every problem is completely independent of all others, so an incorrect answer on one problem typically will not carry over to subsequent problems.

Topics and the approximate distribution of problems on the afternoon session of the civil construction depth exam are as follows.

Earthwork Construction and Layout (10%)

Excavation and embankment (cut and fill); borrow pit volumes; site layout and control; earthwork mass diagrams

Estimating Quantities and Costs (17.5%)

Quantity take-off methods; cost estimating; engineering economics; value engineering and costing

Construction Operations and Methods (15%)

Lifting and rigging; crane selection, erection, and stability; dewatering and pumping; equipment production; productivity analysis and improvement; temporary erosion control

Scheduling (17.5%)

Construction sequencing; CPM network analysis; activity time analysis; resource scheduling; time-cost trade-off

Material Quality Control and Production (10%)

Material testing (e.g., concrete, soil, asphalt); welding and bolting testing; quality control process (QA/QC); concrete mix design

Temporary Structures (12.5%)

Construction loads; formwork; falsework and scaffolding; shoring and reshoring; concrete maturity and early strength evaluation; bracing; anchorage; cofferdams (systems for temporary excavation support); codes and standards (e.g., American Society of Civil Engineers (ASCE 37), American Concrete Institute (ACI 347), American Forest & Paper Association (NDS), Masonry Wall Bracing Standard)

Worker Health, Safety, and Environment (7.5%)

OSHA regulations; safety management; safety statistics (e.g., incident rate, EMR)

Other Topics (10%)

Groundwater and Well Fields—groundwater control including drainage, construction dewatering

Subsurface Exploration and Sampling—drilling and sampling procedures

Earth Retaining Structures—mechanically stabilized earth wall, soil and rock anchors

Deep Foundations—pile load test, pile installation

Loadings—wind loads, snow loads, load paths

Mechanics of Materials—progressive collapse

Materials—concrete (prestressed, post-tensioned), timber

Traffic Safety—work zone safety

For further information and tips about the civil engineering PE exam, consult the *Civil Engineering Reference Manual*, or PPI's website, **www.ppi2pass.com**.

THIS BOOK'S ORGANIZATION

Six-Minute Solutions for Civil PE Exam Construction Problems is organized into two sections. The first section, Breadth Problems, presents 20 construction engineering problems of the type that would be expected in the morning part of the civil engineering PE exam. The second section, Depth Problems, presents 80 problems representative of the afternoon part of this exam. The two sections of the book are further subdivided into the topic areas covered by the construction exam.

Most of the problems are quantitative, requiring calculations to arrive at a correct solution. A few are non-quantitative. Some problems will require a little more than six minutes to answer and others a little less. On average, you should expect to complete 80 problems in 480 minutes (eight hours), or spend six minutes per problem.

HOW TO USE THIS BOOK

Six-Minute Solutions for Civil PE Exam Construction Problems presents problems in the same format as those encountered on the PE exam. The solutions are presented in a step-by-step sequence to help you follow the logical development of the correct solution and to provide examples of how you may want to approach your solutions as you take the PE exam.

Each problem includes a hint to provide direction in solving the problem. In addition to the correct solution, you will find an explanation of the faulty solutions leading to the three incorrect answer options. The incorrect options represent answers derived from common mistakes made when solving each type of problem. These may be simple mathematical errors, such as failing to square a term in an equation, or more serious errors, such as using the wrong equation.

Although problems identical to those presented in *Six-Minute Solutions for Civil PE Exam Construction Problems* will not be encountered on the PE exam, the approach to solving problems will be similar. Solutions presented for each example problem may represent only one of several methods for obtaining a correct answer. Although most of these problems have unique solutions, alternative problem-solving methods may produce a different, but nonetheless appropriate, answer.

The following suggestions are provided to help you optimize your preparation time and obtain the maximum benefit from the practice problems.

1. Complete an overall review of the problems, and identify the subjects that you are least familiar with. Work a few of these problems to assess your general understanding of the subjects and to identify your strengths and weaknesses.

2. Locate and organize relevant resource materials. (See the References section of this book for guidance.) As you work your problems, some of these resources will emerge as more useful to you than others. These are what you will want to have on hand when taking the PE exam.

3. Work the problems in one subject area at a time, starting with the subject areas that you have the most difficulty with.

4. When possible, work problems without utilizing the hint. Always attempt your own solution before looking at the solutions provided in the book. Use the solutions to check your work or to provide guidance in solving the more difficult problems. Use the incorrect answer options to help identify pitfalls and to develop strategies to avoid them.

5. Use each subject area's solutions as a guide to understanding general problem-solving approaches.

References

The minimum recommended library for the civil PE exam consists of PPI's *Civil Engineering Reference Manual.* You may also find the following references helpful in completing problems in *Six-Minute Solutions for Civil PE Exam Construction Problems,* as well as during the exam.

American Concrete Institute. *Formwork for Concrete* (ACI SP-4).

American Concrete Institute. *Guide to Formwork for Concrete* (ACI 347).

American Forest & Paper Association/American Wood Council. *National Design Specification for Wood Construction, ASD/LRFD* (NDS).

American Institute of Steel Construction. *Steel Construction Manual* (AISC).

American Society of Civil Engineers. *Design Loads on Structures During Construction* (ASCE 37).

California Department of Transportation. *Stormwater Pollution Prevention Plan (SWPPP) and Water Control Program (WPCP) Preparation Manual.*

Council for Masonry Wall Bracing, Mason Contractors Association of America. *Standard Practice for Bracing Masonry Walls During Construction* (CMWB).

Fiori, Christine M., Kraig Knutson, Richard Mayo, and Clifford J. Schexnayder. *Construction Management Fundamentals.* McGraw-Hill Science/Engineering/Math.

Harris, Frank. *Modern Construction and Ground Engineering Equipment and Methods.* Longman-Publishing Group.

Hinze, Jimmie W. *Construction Planning and Scheduling.* Prentice Hall.

Ledbetter, William B., Robert L. Peurifoy, and Clifford J. Schexnayder. *Construction Planning, Equipment, and Methods.* McGraw-Hill Companies.

Lessard, Charles, and Joseph Lessard. *Project Management for Engineering Design: Synthesis Lectures on Engineering.* Morgan and Claypool Publishers.

Nunnally, S. W. *Construction Methods and Management.* Prentice Hall.

Ratay, Robert T. *Handbook of Temporary Structures in Construction.* McGraw-Hill Professional.

U.S. Department of Labor. *Occupational Safety and Health Standards for the Construction Industry, 29 CFR Part 1926* (OSHA).

U.S. Federal Highway Administration. *Manual on Uniform Traffic Control Devices—Part 6: Temporary Traffic Control* (MUTCD–Pt 6).

Codes Used to Prepare This Book

The information that was used to write this book was based on the exam-adopted materials at the time of publication. However, as with engineering practice itself, the PE examination is not always based on the most current codes or cutting-edge technology. Similarly, codes, standards, and regulations adopted by state and local agencies often lag issuance by several years. It is likely that the codes that are most current, the codes that you use in practice, and the codes that are the basis for your exam will all be different.

PPI lists on its website the dates and editions of the codes, standards, and regulations on which NCEES has announced the PE exams are based. It is your responsibility to find out which codes will be tested on your exam. In the meantime, here are the codes that have been incorporated into this book.

ACI 318: *Building Code Requirements for Structural Concrete*, American Concrete Institute, 2008.

ACI 347: *Guide to Formwork for Concrete* (in ACI SP-4, Seventh ed. appendix), American Concrete Institute, 2004.

ACI SP-4: *Formwork for Concrete*, Seventh ed., American Concrete Institute, 2005.

AISC: *Steel Construction Manual*, Thirteenth ed., American Institute of Steel Construction, 2005.

ASCE 37: *Design Loads on Structures During Construction*, American Society of Civil Engineers, 2002.

CMWB: *Standard Practice for Bracing Masonry Walls During Construction*, Council for Masonry Wall Bracing, Mason Contractors Association of America, 2001.

MUTCD–Pt 6: *Manual on Uniform Traffic Control Devices–Part 6 Temporary Traffic Control*, U.S. Federal Highway Administration, 2009.

NDS: *National Design Specification for Wood Construction ASD/LRFD*, American Forest & Paper Association/American Wood Council, 2005.

OSHA: *Occupational Safety and Health Standards for the Construction Industry*, 29 CFR Part 1926, (U.S. federal version), U.S. Department of Labor.

Nomenclature

a	depth	in
a	length	ft
A	area	ft^2
A	contribution	$
A	gradient change	%
A	monthly payment value	$
ACWP	actual cost of work performed	$
b	length	ft
b	width	yd
b_w	cross-section width	in
B	depth	ft
BAC	budget at completion	$
BCWP	budgeted cost of work performed	$
BCY	bank cubic yards or bank volume	yd^3
BS	backsight	ft
c	cost	$
c	depth of Whitney stress block	in
C	coefficient	–
C	cost	$
CCY	compact cubic yards or compacted soil volume	yd^3
CPI	cost performance index	–
d	depth	ft
d	distance	ft
D	dead load	lbf or lbf/ft
D	depreciation	$
D	diameter	various
D	duration	day
DF	shrinkage	%
E	energy	ft-lbf
E'	modulus of elasticity	lbf/in^2
EAC	estimate at completion	$
EF	earliest finish	day
ES	earliest start	day
ETC	estimated cost to completion	$
EVM	earned value management	–
f	stress	in
f'_c	concrete compressive strength	lbf/in^2
f_y	steel tension yield strength	lbf/in^2
F	force	lbf
F	future worth	$
F'	allowable stress	lbf/in^2
FF	free float	day
FS	factor of safety	–
FS	foresight	ft
g	gravitational acceleration, 32.2	ft/sec^2
g	gravity	ft/sec^2
g_c	gravitational constant, 32.2	lbm-ft/lbf-sec^2
h	depth	ft
h	effective embedment of anchor	in
h	hydraulic head	ft
h'	height at top of strut	ft
H	height	in
H	lateral load at top of form	lbf/ft
i	hydraulic gradient	in/in
i	monthly interest rate	%
I	importance factor	–
I	intersection angle	deg
I	moment of inertia	in^4
Ib/Q	rolling shear constant	in^2
k_a	active lateral earth pressure coefficient	–
K	basic concrete breakout strength in tension	lbf
K	coefficient of permeability	in/sec
K	coefficient of restitution	–
l	length	ft
l'	horizontal distance from form to bottom of strut	ft
L	concentrated live load	kips
L	lag time	day
L	length	ft
LCY	loose bank volume	yd^3
LF	latest finish	day
LH	labor hour	$/hr
LS	latest start	day
m	mass	lbm
$\dot{m}$	mass flow rate	gal/min
n	length	ft
n	quantity	–
N	basic concrete breakout strength	lbf
p'	axial load	lbf/ft
P	present value	$
P	production rate	$\text{yd}^3\text{/hr}$
P	total load	lbf
q	lateral wind pressure	lbf/ft^2
Q	quantity of flow	$\text{in}^3\text{/sec}$
r	radius	ft
R	lateral force	lbf
R	radius	ft
R	rate of change in grade per station	%/sta
R	rate of placement	ft/hr
R	safe driving load	lbf
s	length of pile side	in
S	average blow penetration	in
S	section modulus	in^3
S	speed limit	mph
SF	swell factor	%
SG	specific gravity	–
SPI	schedule performance index	–
SV	schedule variance	$
t	thickness	in

t	time	hr
T	temperature	°F
T	tensile force	lbf
T	tension	lbf
TF	total float	day
v	velocity	ft/sec
V	volume	ft³
$\dot{V}$	volumetric flow rate	ft³/sec
w	density of concrete	lbf/ft³
w	width	ft
w	wind load	lbf/ft²
W	weight	lbf or lbf/ft
WHP	water (hydraulic) horsepower	hp
x	amount of product	–
x	distance to horizontal center of gravity	ft
y	depth of water	ft
y	profit	$

Symbols

α	deflection angle	deg
β	half of the angle that an arc encloses	deg
β_1	ratio of depth of equivalent stress block to depth of actual neutral axis	–
ϵ	tensile steel strain	–
λ	modification factor	–
γ	unit or specific weight	lbf/ft³

μ	mean	various
ϕ	angle of repose	deg
ϕ	strength reduction factor	–
ψ	anchor tensile strength modification factor	lbf
ρ	density	lbm/ft³
σ	standard deviation	day
σ'	effective stress	lbf/ft²

Subscripts

A	additional	
c	constant or curvature	
d	dry	
D	dynamic	
E	extraction	
F	friction	
i	i event or initial	
m	minor	
p	pressure	
req	required	
s	solids	
trap	trapezoid	
u	uniformity	
v	velocity	
w	water	
x	x-direction	
y	y-direction	
z	elevation	

Breadth Problems

Earthwork Construction and Layout

1. An area is to be excavated with excavation depths as shown. A truck can carry 12 yd³ of soil. The soil's swell factor is 8%. What is most nearly the number of truck loads that will be needed to transport the excavated soil to a dump site?

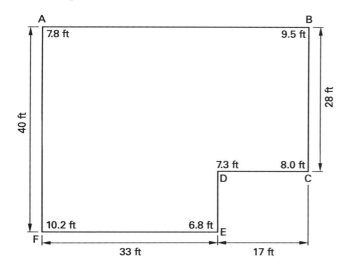

(A) 40 loads

(B) 45 loads

(C) 50 loads

(D) 60 loads

Hint: First find the volume of excavated soil.

2. The diagram shown represents the cross section of a site. The soil's shrinkage factor is 12%. What is most nearly the net cut from sta 2+00 to sta 2+50?

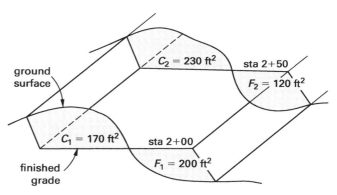

(A) 35 yd³

(B) 75 yd³

(C) 300 yd³

(D) 370 yd³

Hint: The net cut volume takes into account both the cut and fill volumes.

3. Compacted soil 5 ft deep is needed for a building's 200 ft × 100 ft foundation. The soil will be borrowed from a loose bank. The borrow soil has a density of 120 lbm/ft³, a water content of 7%, a shrinkage factor of 10%, and a swell factor of 12%. The required loose bank volume and contained water weight are most nearly

(A) 4120 yd³; 880,000 lbf

(B) 4120 yd³; 940,000 lbf

(C) 4610 yd³; 977,000 lbf

(D) 4610 yd³; 1,050,000 lbf

Hint: Use the soil density to obtain soil weight from soil volume and find the weight of the water.

4. A contractor is obtaining soil fill from a 200 ft $\times$ 150 ft borrow pit. The soil will be excavated to a depth of 3 ft. The swell factor of the soil is 12%, and the specific weight of the soil after excavation will be 115 lbf/ft^3. A single truck will be used to transport all the soil volume to a dump site. The truck's maximum capacity per load is the lesser of 12 yd^3 or 15 tons. Approximately how many loads will be required to transport the soil fill to the dump site?

(A) 310 loads

(B) 350 loads

(C) 380 loads

(D) 390 loads

Hint: Find the maximum volume the truck can carry.

5. What is most nearly the deflection angle of the arc in the horizontal curve shown?

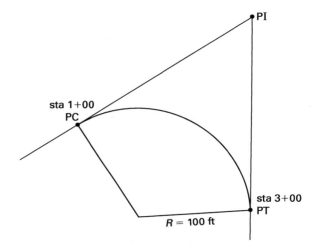

(A) 30°

(B) 35°

(C) 60°

(D) 120°

Hint: The deflection angle is the angle between the tangent and the chord.

6. A road's vertical curve is shown. What is most nearly the elevation at point P?

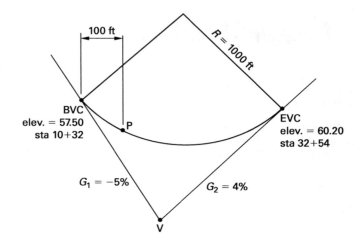

(A) 52.3 ft

(B) 52.5 ft

(C) 52.6 ft

(D) 52.7 ft

Hint: Calculate the elevation difference between the road and the tangent.

Estimating Quantities and Costs

7. The foundation wall and footing shown are located around the perimeter of a building's footprint. Approximately how much concrete will be required for the building's perimeter?

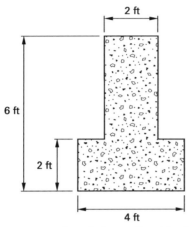

foundation wall and footing

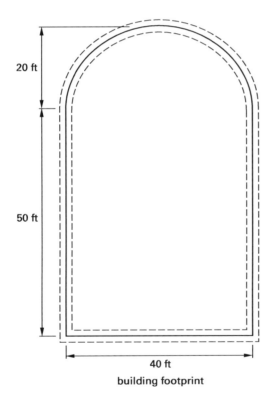

building footprint

(A) 100 yd^3

(B) 120 yd^3

(C) 150 yd^3

(D) 160 yd^3

Hint: Find the foundation's volume of concrete per linear foot, and then find the length of the building's perimeter.

8. A foundation wall and footing are to be formed with 2 in plywood sheathing. Approximately how much form-work in units of board foot measure (bd-ft) is needed for the foundation and the foundation wall?

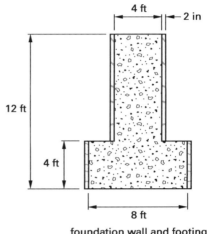

foundation wall and footing

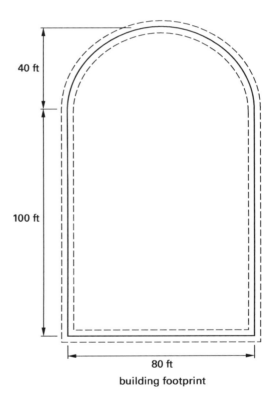

building footprint

(A) 17,300 bd-ft

(B) 19,500 bd-ft

(C) 25,500 bd-ft

(D) 25,900 bd-ft

Hint: Calculate the board foot measure per foot of foundation, and then obtain the length of the building perimeter.

9. One foreman (at $50/hr), two carpenters (at $45/hr each), one painter (at $35/hr), and one laborer (at $25/hr) are working on a residential construction project. The overtime rate for each project member is 50% more than the normal rate. The overtime labor hour for the entire project team is most nearly

(A) $58/hr

(B) $60/hr

(C) $200/hr

(D) $300/hr

Hint: Calculate the crew's average hourly rate.

10. A company's base hourly rate for a welder is $30. The benefits offered by the company are estimated based on a percentage of the base wage: social security, 6%; worker's compensation insurance, 5%; and general liability insurance, 4%. Health and dental insurance is $400 a month, with 75% company assistance. Assume a 30 day month and that the welder works 40 hours per week with no overtime. The weekly cost of employing the welder is most nearly

(A) $1200/wk

(B) $1400/wk

(C) $1500/wk

(D) $1800/wk

Hint: The real hourly rate will include base pay and benefits.

Scheduling

11. What is the correct sequence for the following construction activities?

I.   finishing concrete floor slab

II.   pouring concrete roof

III. conducting excavation

IV. pouring concrete columns

V.   pouring foundation

VI. setting up steel roof joists and composite metal roofing

(A) V, III, IV, I, VI, and II

(B) III, V, I, IV, VI, and II

(C) III, V, IV, I, II, and VI

(D) III, V, IV, I, VI, and II

Hint: Evaluate sequentially.

12. What is the correct sequence for constructing a single-span bridge?

I.   placing hot mixed asphalt as pavement

II.   placing steel stringers and floor beams

III.   pouring concrete deck

IV.   pouring concrete for abutments

V.   setting up reinforcement for deck

VI.   driving foundation piles

VII. placing concrete forms for deck

VIII. building concrete pile caps

(A) VI, VIII, IV, II, VII, V, III, and I

(B) VIII, VI, IV, II, V, VII, III, and I

(C) VI, VIII, IV, V, II, VII, III, and I

(D) VI, VIII, IV, II, V, VII, III, and I

Hint: Evaluate sequentially.

13. The following bar chart represents the construction schedule for a residential house. The owner of the house wishes to visit the construction site when the most construction activities are in progress. Assume a seven day work week. What week should the owner visit the site?

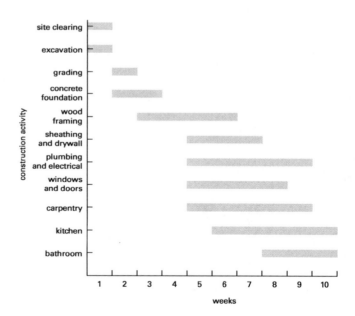

(A) week 5

(B) week 6

(C) week 7

(D) week 8

Hint: Shaded areas represent activities conducted in a given week.

14. Resources required to finish a project on time are allocated as shown. Which week will utilize the most resources if every activity starts at the earliest possible time?

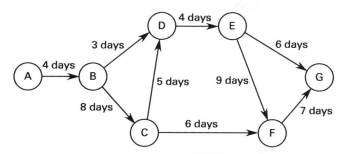

activity	resources (number of people)
AB	7
BC	5
BD	16
CD	11
DE	6
CF	7
EF	6
EG	9
FG	5

(A) week 1

(B) week 2

(C) week 3

(D) week 4

Hint: Determine the earliest start (ES) and earliest finish (EF) of each activity.

15. Which of the following statements about earned value management is INCORRECT?

(A) A CPI greater than or equal to 1.0 suggests positive cash flow.

(B) An SPI less than or equal to 1.0 suggests a project is ahead of schedule.

(C) The SV is measured in dollars, not time.

(D) The BCWS is a spending plan for a project.

Hint: Review terms associated with earned value management.

16. Which of the following statements about time-cost trade-off analysis is INCORRECT?

(A) Normal time is the project duration that has the least cost.

(B) The sequence of critical activities must change when a project is completed in crash time.

(C) A project's cost is the sum of its direct and indirect costs.

(D) Crash time is the shortest amount of time in which a project can be completed.

Hint: Review terms associated with time-cost trade-off analysis.

Material Quality Control and Production

17. According to the Unified Soil Classification System, what classification would be given to a nonplastic soil sample with the following sieve analysis results?

sieve size	soil sample (% passing)
no. 4	97
no. 10	90
no. 40	40
no. 100	8
no. 200	4

(A) SP

(B) SW

(C) SM

(D) SC

Hint: Use the USCS soil classification procedure.

18. Which of the following steps is NOT included in a concrete slump test?

(A) pouring fresh concrete in one-third increments into a hollow metal cone

(B) using a tamping rod to tamp each concrete increment 25 times

(C) letting the concrete sit for three minutes after measuring its height

(D) removing the cone carefully so as not to disturb the concrete

Hint: Review procedures for a concrete slump test.

19. A worker weighs 200 lbf and must carry a 25 lbf piece of equipment while performing a task. The design individual personnel load for this worker is most nearly

(A) 175 lbf

(B) 225 lbf

(C) 250 lbf

(D) 275 lbf

Hint: Consult ASCE 37 Sec. 4.1.1.

20. Which of the following would NOT be considered when calculating the horizontal construction load for a temporary structure?

(A) total vertical loads

(B) personnel loads

(C) equipment reactions

(D) earthquake loads

Hint: Consult ASCE 37 Sec. 4.

Depth Problems

Earthwork Construction and Layout

21. A profile mass diagram is shown. What is most nearly the total cut volume in bank cubic yards (BCY)?

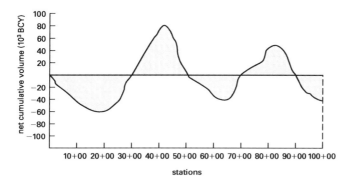

(A) 1.0×10^5 BCY

(B) 1.3×10^5 BCY

(C) 2.3×10^5 BCY

(D) 2.7×10^5 BCY

Hint: Interpret the mass diagram and estimate the total cut volume.

22. A general contractor is excavating soil to build a trench. The trench is 350 ft × 4 ft × 3 ft as shown. The trench's excavated soil is piled into a triangular shape that forms a bank approximately 80 ft long. The angle of repose is 35°, and the soil's swell factor is 18%. What is most nearly the height of the bank?

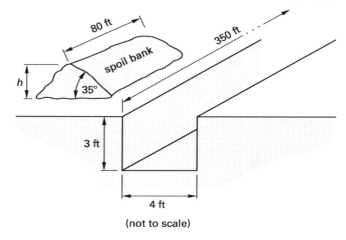

(not to scale)

(A) 4 ft

(B) 5 ft

(C) 6 ft

(D) 7 ft

Hint: Find the base of the triangular bank to calculate its height.

23. 1000 yd³ of compacted soil will be obtained from a borrow pit. The borrow soil's properties are

degree of saturation	70%
water content	15%
specific gravity	2.75

After compaction, the bulk (or total) density of the soil is 115 lbm/ft³, and the optimal moisture content is 18%. Find (a) the volume of soil that must be excavated, and (b) how many gallons of water must be added to the borrow soil.

(A) 900 yd³; 9500 gal

(B) 1000 yd³; 9500 gal

(C) 1040 yd³; 11,300 gal

(D) 1040 yd³; 8230 gal

Hint: Determine the density of the borrow soil to obtain the soil's volume and the gallons of water needed.

24. Soil fill is piled in the form of a conical spoil bank. The cone's diameter is 5 ft, and the soil's angle of repose is 40°. The soil will be used to backfill a 4 ft × 3 ft sloped area that has one side 2 ft taller than the other side. The swell factor is 15%, and the shrinkage factor is 10%. Determine the volume of soil that will either need to be purchased or will be in excess of what is required.

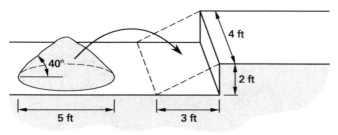

(not to scale)

(A) There is no excess soil, and no soil needs to be purchased.

(B) There is 1.75 ft³ of excess soil, and no soil needs to be purchased.

(C) There is 1.25 ft³ of soil that needs to be purchased.

(D) There is 13.25 ft³ of soil that needs to be purchased.

Hint: Find the volume of the conical shaped pile and convert it to compacted volume.

25. A surveyor has set up ground rods at points A, B, and C on a construction site. The instrument is first placed between point A and point B. At point A, the benchmark elevation is 105.00 ft above mean sea level. The rod at point B gives a foresight reading of 3.85 ft. The rod at point A gives a backsight reading of 5.12 ft. The surveyor then moves the instrument between point B and point C. The rod at point C gives a foresight reading of 6.72 ft. The rod at point B gives a backsight reading of 2.52 ft. What is most nearly the elevation at point C?

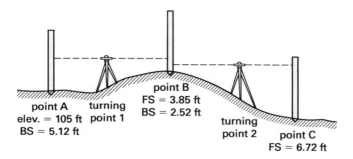

(A) 100 ft

(B) 102 ft

(C) 108 ft

(D) 110 ft

Hint: First obtain the differences in elevations between points.

26. The slope stake shown is located on flat ground. What is most nearly the total cross-sectional area that will be excavated?

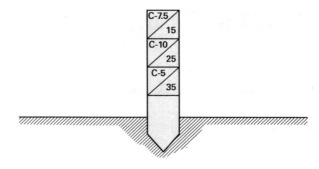

(A) 220 ft²

(B) 380 ft²

(C) 540 ft²

(D) 1070 ft²

Hint: Use the information on the slope stake to calculate the cut area.

27. It costs a contractor $2/yd³ for freehaul and $3/yd³ for overhaul, or $2/yd³ to hire a subcontractor to haul soil off site. The freehaul distance is equal to the average haul distance. Use the mass diagram shown to determine which of the following statements is correct.

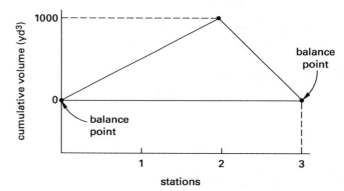

(A) The contractor's haul cost is $500 less than the subcontractor's haul cost.

(B) The subcontractor's haul cost is $500 less than the contractor's haul cost.

(C) The subcontractor's haul cost is $1500 less than the contractor's haul cost.

(D) The subcontractor's haul cost is $2500 less than the contractor's haul cost.

Hint: Calculate average haul distance and obtain cumulative overhaul volume.

28. The profile of a 240 ft long segment of road is shown. The road is 12 ft wide along its entire length. All cut and fill areas are assumed to be perfect semicircles and quarter-circles. The soil's swell factor is 12%. Assume the soil does not shrink in volume when being filled. If a truck has a capacity of 10 yd^3, approximately how many truck loads will it take to transport the soil off site?

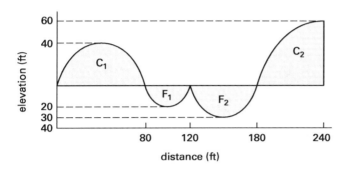

(A) 90 loads

(B) 150 loads

(C) 160 loads

(D) 240 loads

Hint: Calculate the excess volume and convert it to truck loads.

Estimating Quantities and Costs

29. The steel frame for a building in plan and elevation is shown. The columns are W12 × 24, the beams are W14 × 30, and the joists are W12 × 26. What is most nearly the amount of steel needed for the frame?

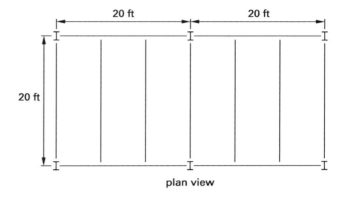

plan view

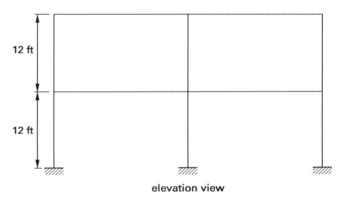

elevation view

(A) 4.0 tons

(B) 7.4 tons

(C) 8.0 tons

(D) 9.2 tons

Hint: Calculate the weight of the joists, the beams, and the columns separately, then combine the three to get the total weight.

30. A house's perimeter is shown. The house has a front and a back door, each 8 ft × 3 ft, and has 12 windows, each 4 ft × 3 ft. The four 15 ft tall brick walls are to be constructed using $3^3/4$ in × $2^1/4$ in × 8 in bricks, and $^1/2$ in thick mortar will be used between bricks. What is most nearly the number of bricks that will be needed to construct the walls?

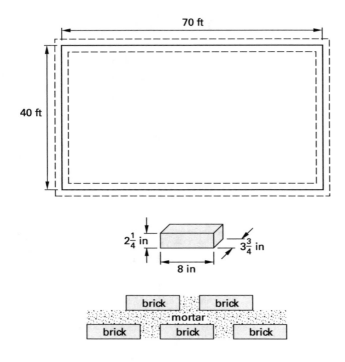

(A) 15,000 bricks

(B) 19,000 bricks

(C) 20,000 bricks

(D) 25,000 bricks

Hint: Obtain the total surface area for the walls.

31. Eight masonry partition walls will be used in a warehouse. Each masonry wall will be 15 ft wide and 12 ft tall and will include

    two windows, 3 ft × 4 ft
    four mechanical conduit openings, circular
       with 1 ft diameter
    one louver, 1.5 ft × 1.5 ft

The bricks are $3^5/8$ in × $2^1/4$ in × $7^5/8$ in. $^1/2$ in thick mortar will be used between bricks. What is most nearly the amount of mortar that will be needed?

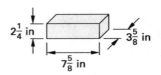
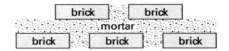

(A) 0.5 yd³

(B) 3.0 yd³

(C) 3.5 yd³

(D) 3.7 yd³

Hint: Multiply the volume of mortar per brick by the total number of bricks.

32. The cross-sectional area for a 100 ft long retaining wall with vertical and horizontal reinforcements is shown. The unit weight of steel is 490 lbf/ft³. What is most nearly the weight of reinforcement needed for the retaining wall?

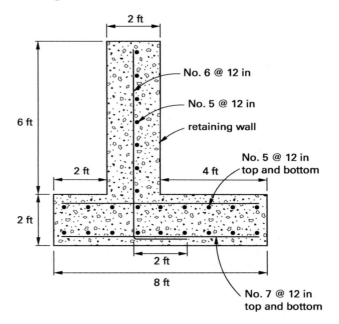

(A) 5000 lbf

(B) 6800 lbf

(C) 7100 lbf

(D) 7400 lbf

Hint: Find the transverse reinforcement and longitudinal reinforcement, and then multiply by the appropriate rebar unit weight.

33. The elevation and cross-sectional views of a small truss bridge are shown. Steel member sizes are as follows: top chord, W18 × 50; bottom chord, W24 × 68; vertical members, C15 × 50; and diagonal members,

W21 × 48. The floor beams are W24 × 104 steel members spaced 10 ft apart, and the concrete deck is constructed using normal weight concrete. What is the approximate dead load on one of the abutments?

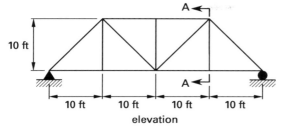

elevation

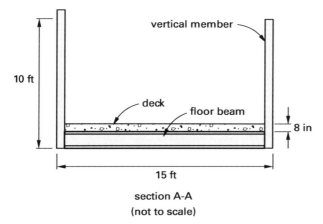

section A-A
(not to scale)

(A) 12,000 lbf

(B) 39,000 lbf

(C) 42,000 lbf

(D) 84,000 lbf

Hint: Calculate the total load for each type of member, and then distribute the load to the two abutments.

34. At the start of a project, a client requests that the construction of a 10,000 ft^2 roof be completed sooner than originally proposed. The existing construction team consists of six people with a labor hour of $45 and a productivity rate of 32 ft^2/hr. Work days are 8 hours and overtime is prohibited. Which of the following is the most cost-effective option for completing the project ahead of schedule?

(A) Use the existing project team.

(B) Add two junior contractors (at $20/hr) to the project team to increase the overall productivity rate to 40 ft^2/hr.

(C) Add two senior contractors (at $50/hr) to the project team to increase the overall productivity rate to 50 ft^2/hr.

(D) Contract the job out to a subcontractor who employs a 10-member project team that has a labor hour of $50 and can finish the job in 20 days.

Hint: Compare total costs.

35. A general contractor is applying for a five year loan to purchase new construction equipment. The maximum monthly payment the contractor can afford to make is $3000, and the loan interest rate is 8%. The maximum amount of money the contractor can borrow is most nearly

(A) $40,000

(B) $150,000

(C) $180,000

(D) $220,000

Hint: Determine the present value based on the monthly payment and interest rate.

36. 10 concrete pads will be constructed with the dimensions shown. Materials costs for this project are

concrete	$200/$yd^3$, poured at 1 yd^3/hr
reinforcement	$3/lbf, built at 90 lbf/hr
formwork	$6/$ft^2$, constructed at 7 ft^2/hr

The project team consists of two concrete workers (at $55/hr), three ironworkers (at $60/hr), and three carpenters (at $40/hr). What is the approximate total cost of material and labor for the construction of the 10 pads?

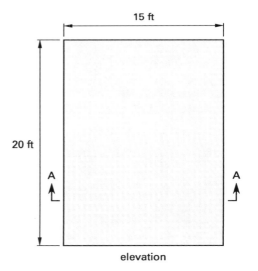

elevation

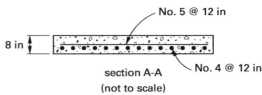

section A-A
(not to scale)

(A) $6100

(B) $44,000

(C) $61,000

(D) $450,000

Hint: Determine the construction completion time to obtain the cost of labor on top of material cost.

37. A construction team is constructing a steel building frame. The team consists of one site engineer (at \$100/hr), one foreman (at \$70/hr), four ironworkers (at \$65/hr each), and one crane operator (at \$50/hr). The frame's components include

steel joists	W16 × 36, 20 ft each, 250 total
steel beams	W24 × 55, 25 ft each, 200 total
steel columns	W14 × 38, 15 ft each, 100 total
steel cost	\$2/lbf
steel fabrication	\$0.5/lbf

Additionally, a crane must be rented at \$2000 per day, and the connections, welding, and splices are estimated to be 15% of the steel cost and fabrication. If 35 pieces of steel can be erected per day, the total cost of the steel frame construction is most nearly

(A) \$1,370,000

(B) \$1,510,000

(C) \$1,540,000

(D) \$1,570,000

Hint: Calculate the weight of the steel to obtain the material cost. Then add on the estimated labor and equipment cost.

38. A new office as shown will include four brick partitions, each with a 3 ft × 4 ft window.

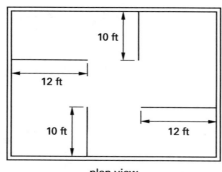

plan view

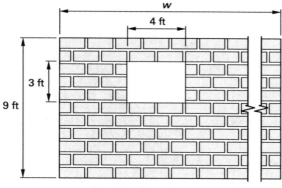

elevation view of partition wall
(not to scale)

The mortar for the bricks will take up approximately 5% of the total brick area, and bricks may be bought at \$10 per square foot. Ignore the price of mortar. The construction team consists of one brick mason (at \$65/hr) and two assistants (at \$40/hr each). The productivity rate of the team is 2 ft²/hr. The same brick mason will work with only one assistant to repoint the mortar to fix imperfections. The repointing productivity rate is 3 ft²/hr. The team will not work overtime. The total cost of constructing the brick partitions is most nearly

(A) \$17,000

(B) \$18,000

(C) \$39,000

(D) \$41,000

Hint: Obtain the total area of brick wall, and then calculate the cost of brick and all labors.

39. A foundation for a new building is to be excavated as shown.

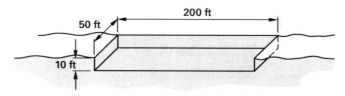

Soil will be excavated at \$12/yd³ and hauled away at \$150 per trip by a truck with an 18 yd³ capacity. Sheet piles, supported by steel wales and struts both made from W24 × 76 steel beams, will be driven 2 ft below excavation and 3 ft above ground at \$13/ft². It is estimated that 1000 ft of steel wales and struts at \$2.5/lbf will be needed to properly support the sheet piles. The project is divided into two crews. Crew 1 will drive the sheet piles and consists of

> two piledriver operators (at \$60/hr)
> two assistants (at \$30/hr each)
> equipment rental (at \$1200/day)
> productivity rate (at 50 ft²/hr)

Crew 2 will place the steel wales and struts and consists of

> one ironworker (at \$50/hr)
> two laborers (at \$30/hr)
> productivity rate (at 20 ft/hr)

The total cost for constructing the building's foundation is most nearly

(A) \$230,000

(B) \$370,000

(C) \$390,000

(D) \$420,000

Hint: Start by finding the volume of soil to be excavated.

40. A contractor plans to buy a sheepsfoot roller for \$20,000 and use it for five years. The salvage value of the roller is \$5000. Using the sum-of-the-years' digits method, what is the approximate worth of the equipment after the end of its third year of use?

(A) $5000

(B) $8000

(C) $10,000

(D) $11,000

Hint: Estimate the depreciation each year according to the previous year's worth.

41. A 30 year old engineer's annual salary is $70,000. 5% of the engineer's monthly income is deposited into a 401(k) savings account that has an average interest rate of 8% per year. If the engineer never receives a raise, approximately how many years must the engineer work in order to retire with $1,000,000 in the 401(k)?

(A) 7 yr

(B) 15 yr

(C) 40 yr

(D) 70 yr

Hint: Calculate the number of monthly payments required to reach the desired future value.

42. A project's progress was tracked in the graph shown. Which statement is a correct assessment about the project's performance?

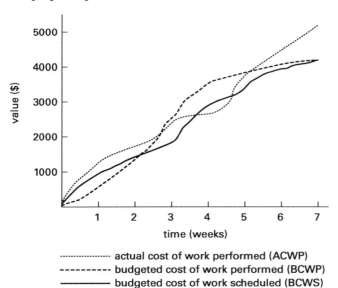

----------- actual cost of work performed (ACWP)
--------- budgeted cost of work performed (BCWP)
————— budgeted cost of work scheduled (BCWS)

(A) The project became profitable as it moved toward completion.

(B) The project was profitable from week three to week five.

(C) The project was behind schedule for its entire duration.

(D) The cost performance index for the completed project was greater than one.

Hint: Interpret the graph using earned value management (EVM) definitions.

Construction Operations and Methods

43. Two slings are arranged asymmetrically to lift three objects as shown. Object A weighs 3000 lbf, and object B weighs 1500 lbf. The mass of the spreader beam is negligible. When the slings are pulled up at the point where they meet, the three objects are raised and remain horizontally aligned. What does object C most nearly weigh?

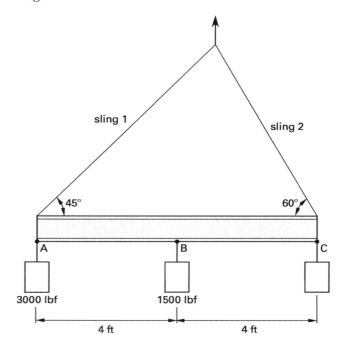

(A) 3000 lbf

(B) 5300 lbf

(C) 5700 lbf

(D) 7500 lbf

Hint: Find the weight of object C using principles of static equilibrium.

44. Two different materials make up the object shown. Material B weighs 5000 lbf, and material C weighs 10,000 lbf. The master link is properly placed. What is most nearly the length of sling 1?

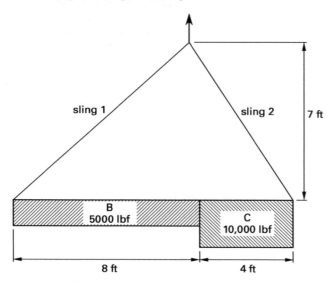

(A) 8.1 ft

(B) 9.2 ft

(C) 11 ft

(D) 19 ft

Hint: Find the horizontal center of gravity of the object.

45. A tower crane is equipped with a 246 ft boom, as well as load handling accessories (e.g., slings and spreader bars) that have a combined weight of 2000 lbf. The crane's load capacities are given in the table provided. What is the approximate maximum safe lift radius for a 15,000 lbf load?

lift radius (ft)	boom length (ft)	
	246	265
	load capacity (lbf)	
70	44,090	41,190
93	31,990	29,280
110	26,230	23,960
120	23,850	21,760
131	21,720	19,780
140	20,020	18,200
150	18,440	16,730
170	15,940	14,420
180	14,750	13,320
190	13,810	12,440
208	12,280	11,020
220	11,380	9570
230	10,690	8640
246	9700	7720

(A) 150 ft

(B) 160 ft

(C) 170 ft

(D) 250 ft

Hint: Add the weight of the load handling accessories to the load, and then interpolate the table to find the maximum radius.

46. The mobile crane shown has a cab weight of 28 tons and a boom weight of 3 tons. The crane must lift a 5 ton steel truss. A safety factor of 1.5 will be used. What is the required minimum tonnage of the counterweight, if any, to safely lift the truss?

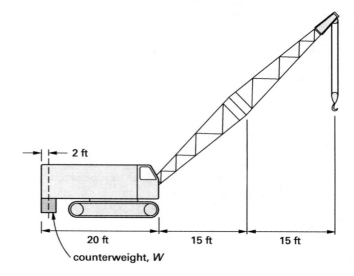

counterweight, *W*

(A) The crane requires a counterweight of 0.7 tons.

(B) The crane requires a counterweight of 5 tons.

(C) The crane will not require a counterweight because the weight of the crane cab is greater than 1.5 times the weight of the steel truss.

(D) The crane will not require a counterweight because the counter moment is greater than 1.5 times the turning moment.

Hint: The counter moment must be at least 1.5 times the turning moment.

47. An unconfined aquifer is shown with the original water table 200 ft from its bottom. To access the water, a well with a 1 ft radius is drilled. The pumping rate is 80 gpm, and the hydraulic conductivity is 1.0 ft/day. The well drawdown is zero at a distance of 1500 ft from the well. After pumping has continued long enough to reach steady-state conditions, what will be the approximate depth of water in the well?

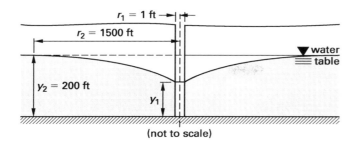

(not to scale)

(A) 0 ft

(B) 30 ft

(C) 65 ft

(D) 90 ft

Hint: Use the Dupuit equation.

48. A pump is used to transfer water from one reservoir to another as shown. The roughness of the pipe causes a head loss of 5 ft in the suction pipe and 7 ft in the discharge pipe. The pump is delivering water at a rate of 500 gpm and maintains a pressure head of 30 psig with a velocity of 1 ft/sec. Friction causes a pressure head loss of 15 psig at the pump inlet and 10 psig at the outlet. Minor losses may be neglected. At what approximate water (hydraulic) horsepower must the pump operate in order to overcome the losses between the two reservoirs?

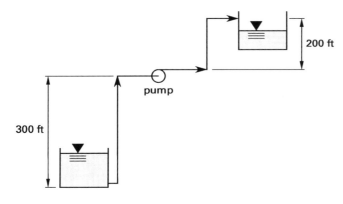

(A) 2.0 hp

(B) 60 hp

(C) 70 hp

(D) 520 hp

Hint: The total head is the sum of static suction head, pressure head, and friction head.

49. A hydraulic excavator is operated for 50 minutes per hour to excavate soft clay at a job site. The excavator has a bucket capacity of 2 yd^3 and a 90° swing angle. The bucket factor is 0.9. The excavator can cut a maximum depth of 20 ft. At the job site, the excavator's average cut is 10 ft. Using Table 49.1 and Table 49.2, what is most nearly the hourly anticipated production of the excavator?

Table 49.1: Machine Performance

	machine size (cycles per hr)			
material type	wheel tractor	small excavator (≤ 1 yd)	medium excavator ($1^{1}/_{4}$–$2^{1}/_{4}$ yd)	large excavator ($> 2^{1}/_{2}$ yd)
soft (sand, gravel, loam)	170	250	200	150
average (common earth, soft clay)	135	200	160	120
hard (tough clay, rock)	110	160	130	100

Table 49.2: Swing Depth Factor

cut depth (% of maximum)	angle of swing (deg)					
	45	60	75	90	120	180
30	1.33	1.26	1.21	1.15	1.08	0.95
50	1.28	1.21	1.16	1.10	1.03	0.91
70	1.16	1.10	1.05	1.00	0.94	0.83
90	1.04	1.00	0.95	0.90	0.85	0.75

(A) 240 yd^3/hr

(B) 260 yd^3/hr

(C) 290 yd^3/hr

(D) 320 yd^3/hr

Hint: Calculate the volume of earth the excavator can remove in an hour.

50. A 200 ft × 300 ft backfill site will be compacted by a 15 ton static roller. The roller width is 8 ft, and the site will be rolled twice for each lift. The lift depth is 24 in. If the roller operates at a speed of 5 mi/hr, approximately how many hours are required to finish two lifts?

(A) 0.95 hr

(B) 0.28 hr

(C) 0.57 hr

(D) 1.2 hr

Hint: Calculate the production per hour for one lift.

51. The cycle time of a scraper is 5 min. The cycle times for single and tandem pushers are given. How many pushers would be required to serve a fleet of eight scrapers if a single pusher and the chain method are used?

loading method	single pusher	tandem pusher
back-track	1.5	1.4
chain or shuttle	1.0	0.9

(A) 1

(B) 2

(C) 5

(D) 8

Hint: Calculate the number of scrapers each pusher can handle.

52. A given scraper has a cycle time of 2 min, excluding the time to load material onto the scraper. The scraper is observed, and load time data are collected as shown. What is most nearly the optimal total cycle time for the scraper so that loading efficiency is optimized?

average load time (min)	soil loaded (yd³)
0.2	10
0.4	18
0.6	24
0.8	28
1.0	30
1.2	31
1.4	31.5

(A) 1.0 min

(B) 1.4 min

(C) 3.0 min

(D) 3.5 min

Hint: The optimal load time can be found by plotting a graph of load growth data.

53. Which of the following are NOT used during excavation as a temporary erosion control measure?

(A) sheet piles

(B) silt fences

(C) slope drains

(D) brush barriers

Hint: Review temporary erosion control measures.

54. A construction project requires the installation of an irrigation pipe that is 200 ft long and 1 ft in diameter. A trench 6 ft deep × 3 ft wide × 200 ft long is dug for the pipe, and this is backfilled with borrow pit soil. The support system consists of vertical wood studs, horizontal wood wales, and cross bracing that spans in between the wales, as shown. A 12 in deep gravel base is underneath the irrigation pipe. The cost per unit for materials is provided. What is most nearly the total combined cost of these materials?

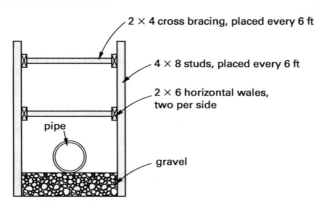

material	cost per unit
timber	$4/bd-ft
gravel base	$36/yd³
borrow pit soil	$25/yd³
pipe	$30/ft

(A) $13,500

(B) $17,500

(C) $17,700

(D) $18,100

Hint: The total material cost includes pipe, fill, and timber.

Scheduling

55. What is the critical path of the network diagram shown?

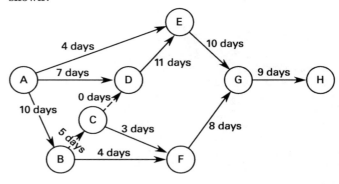

(A) A-D-E-G-H

(B) A-B-F-G-H

(C) A-B-C-F-G-H

(D) A-B-C-D-E-G-H

Hint: Activities on the critical path have zero float.

56. From the network diagram shown, what is the total float for activity BD?

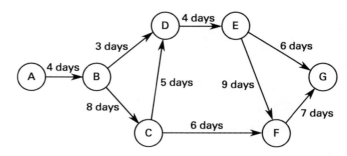

(A) 0 days

(B) 3 days

(C) 10 days

(D) 12 days

Hint: Find the earliest start, earliest finish, latest start, and latest finish time.

57. A project is depicted by the precedence diagram shown. What is the critical path?

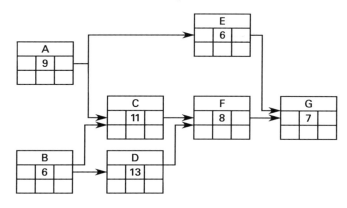

key

ES = earliest start
EF = earliest finish
D = duration
LS = latest start
LF = latest finish
TF = total float

activity description		
ES	D	LS
EF	TF	LF

(A) A-E-G

(B) A-C-F-G

(C) B-C-F-G

(D) B-D-F-G

Hint: The total float of the critical path is always zero.

58. An activity has an earliest start on day 15, an earliest finish on day 28, and a 2 day lag time from finish to start of the successor. If the earliest finish of the predecessor is on day 13 and the earliest start of the successor is on day 34, what is the free float of this activity?

(A) 0 days

(B) 2 days

(C) 4 days

(D) 6 days

Hint: Account for lag time.

59. Which of the following statements can be inferred from the activity-on-node diagram shown?

(A) The duration of activity A is 2 days.

(B) The duration of activity B is 8 days.

(C) The earliest start for activity A is 2 days.

(D) The earliest finish for activity B is 6 days.

Hint: Review activity-on-node diagrams.

60. Using the precedence table given, what is the critical path?

activity	duration (days)	predecessors
A, start	3	–
B	5	A
C	6	A
D	3	B
E	6	B, C
F	7	D, E
G	2	C
H, finish	3	F, G

(A) A-C-G-H

(B) A-B-D-F-H

(C) A-B-E-F-H

(D) A-C-E-F-H

Hint: Draw and complete an activity-on-node CPM network.

61. What type of relationship is shown?

(A) start-to-finish

(B) finish-to-start

(C) start-to-start

(D) finish-to-finish

Hint: Review precedence diagrams.

62. Which of the following would NOT be considered a fixed cost for a construction firm?

(A) rent

(B) utilities

(C) administrative staff

(D) rental equipment

Hint: Fixed costs are also known as indirect costs.

63. A precedence diagram is shown. Which activity has the greatest free float?

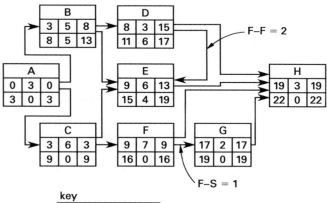

key

ES = earliest start
EF = earliest finish
 D = duration
LS = latest start
LF = latest finish
TF = total float
F–F = finish to start
F–S = start to start

activity description		
ES	D	LS
EF	TF	LF

(A) activity B

(B) activity D

(C) activity E

(D) activity F

Hint: Free float is calculated by subtracting the earliest finish of an activity from the earliest start of its successor activity.

64. A project has a normal completion time of 25 days and will cost $64,000. The project owner would like to reduce the duration of the project and has a maximum budget of $72,000. The project's cost will increase 10% for each day it finishes ahead of its normal completion time; however, the project duration may not be shorter than 20 days due to limited resources. What is most nearly the cost of the project per day if completed under crash time?

(A) $2600/day

(B) $2800/day

(C) $3100/day

(D) $3300/day

Hint: Find the number of days the project can be reduced under crash time.

65. Which equation calculates the projected total final cost of a project upon completion?

(A) $ETC = BAC - BCWP$

(B) $EAC = ACWP + ETC$

(C) $ETC = BAC + BCWP$

(D) $EAC = ACWP - ETC$

Hint: A project's projected total final cost upon completion is known as its estimate at completion.

66. A project manager is trying to determine the number of days a project's estimated completion time may vary from its critical path. From the information given, approximately how many days will it take to complete the project?

activity mean time (days)	activity standard deviation (days)
16	1.5
11	0.7
9	3.1

(A) 36 days

(B) 38 days

(C) 40 days

(D) 48 days

Hint: The variance is the square of the standard deviation.

67. A masonry company manufactures two products that use the same clay, of which there is a maximum amount of 280 ft^3 available per week. Product 1's profit is $140 per ton, and each ton requires 20 ft^3 of clay. Product 2's profit is $160 per ton, and each ton requires 40 ft^3 of clay. The blending machine must be run for five hours per ton of either product 1 or product 2, and an operator may only work up to 50 hours per week. Additionally, the curing vats for product 1 and product 2 are 8 tons and 6 tons, respectively. How many tons of each product should be produced each week to obtain maximum profit?

(A) product 1: 0 tons; product 2: 7 tons

(B) product 1: 2 tons; product 2: 6 tons

(C) product 1: 6 tons; product 2: 4 tons

(D) product 1: 8 tons; product 2: 6 tons

Hint: Use linear programming.

68. The graph shown represents the cost and duration of a project's critical path activities. What is the most cost-effective way to shorten the overall project duration by three days?

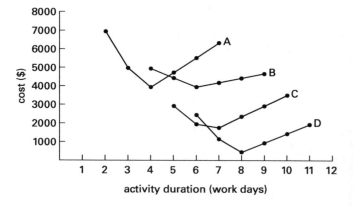

(A) Shorten activity B by three days.

(B) Shorten activity C by two days, and reduce the duration of activity B by one day.

(C) Shorten activity C by one day, and reduce the duration of activity B by two days.

(D) Shorten activities B, C, and D each by 1 day.

Hint: Calculate the cost increase for each activity when time is shortened.

Material Quality Control and Production

69. Half of a shipment of steel reinforcements arrives at a project site lightly coated with steel mill scale and mud. The construction manager should

(A) reject the entire shipment from the reinforcement supplier

(B) reject only the reinforcements coated with mill scale and mud

(C) sand blast the surfaces of the coated reinforcements before placing them in formwork

(D) do nothing because the reinforcements are fine as they are

Hint: Consult ACI 318 Chap. 7.

70. A soil sample has a permeability of 5×10^{-6} in/sec and will be tested using the setup shown. If the pipe's diameter is 2 in, what is most nearly the volume of flow per second?

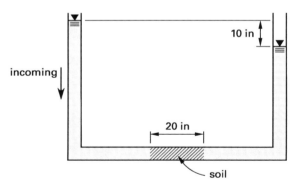

(A) 2.5×10^{-6} in^3/sec

(B) 5.0×10^{-6} in^3/sec

(C) 8.0×10^{-6} in^3/sec

(D) 3.0×10^{-5} in^3/sec

Hint: Use Darcy's law to determine the velocity and volume of flow.

71. What are the lengths of the fillet welds shown?

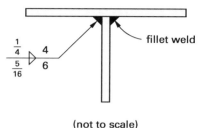

(not to scale)

(A) fillet weld: $5/16$ in on the left and $1/4$ in on the right; weld: 6 in on the left and 4 in on the right

(B) fillet weld: $1/4$ in on the left and $5/16$ in on the right; weld: 6 in on the left and 4 in on the right

(C) fillet weld: $5/16$ in on the left and $1/4$ in on the right; weld: 4 in on the left and 6 in on the right

(D) fillet weld: $1/4$ in on the left and $5/16$ in on the right; weld: 4 in on the left and 6 in on the right

Hint: Review the *AISC Manual's* welding symbols.

72. Which of the following does NOT affect the allowable tensile strength of a single anchor bolt?

(A) bolt length

(B) tensile force magnitude

(C) concrete compressive yield strength

(D) concrete friction angle

Hint: Consult ACI 318 App. D.

73. Quality control involves all of the following EXCEPT

(A) testing

(B) inspecting

(C) documenting

(D) auditing

Hint: Quality control is not the same as quality assurance.

74. A concrete mixture with the following specifications has a cement, sand, aggregate ratio of 1:2.5:3.5 (by weight).

specific gravity of cement	3.5
specific gravity of sand	2.7
specific gravity of aggregate	2.8
no. sacks of cement per yd^3 of concrete	6
weight of one sack	94 lbf
air content voids	0%

What is the mixture's water-cement ratio?

- (A) 1:1.91
- (B) 1:4.76
- (C) 1:6.67
- (D) 1:14.3

Hint: Use the absolute volume method.

75. A batch of concrete must meet the following criteria.

water-cement ratio	1:2.5
air content	5%
fine aggregate	specific gravity, 2.5; moisture content, 5%
coarse aggregate	specific gravity, 3.0; moisture content, 2%
5.5 sacks of cement	specific gravity, 2.7
weight of 1 sack	94 lbf

The concrete contains 50% fine aggregates and 50% coarse aggregates (by volume). To meet the criteria, approximately how much water must be added to 1.0 yd^3 of cement?

- (A) 90 lbf
- (B) 95 lbf
- (C) 100 lbf
- (D) 200 lbf

Hint: Calculate the weight and volume of each component.

Temporary Structures

76. The formwork for a square concrete column will be used at a jobsite for two weeks. The local wind load at the jobsite is 110 mph for an average one-minute gust. If the wind direction, velocity pressure, and topographic factors all have a value of 1.0, what is most nearly the tensile force in brace B?

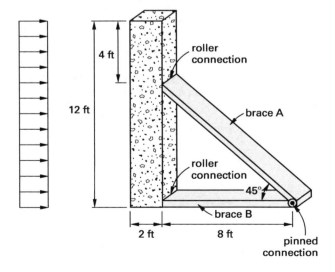

- (A) 100 lbf
- (B) 160 lbf
- (C) 290 lbf
- (D) 690 lbf

Hint: Consult ASCE 37 Sec. 6.2.

77. A construction project calls for concrete walls 13 ft in height. The concrete mixture is made with type I cement, is poured into the wall form at a rate of 5 ft/hr, and has a temperature of 60°F during pouring. The concrete is normal weight and has a unit weight of 150 lbf/ft^3 without any admixture or fly ash. (a) What is most nearly the maximum pressure generated from the wet concrete, and (b) where does it occur?

- (A) 150 lbf/in^2; constant along formwork
- (B) 900 lbf/in^2; bottom of formwork up to 6 ft deep
- (C) 900 lbf/in^2; bottom of formwork
- (D) 1950 lbf/in^2; bottom of formwork

Hint: Use ASCE 37 Eq. 4-2.

78. A long concrete exterior wall is shown. The concrete is poured at a rate of 5 ft/hr at a temperature of 75°F. The deflection limitation is 1/360 of any dimension and there will be at least three spans of formwork. Class I plyform that is 1 in thick and has the following properties is used.

face grain	parallel to support
F_b	1545 lbf/in^2 (ACI SP-4 Table 4-2)
F_v	57 lbf/in^2 (ACI SP-4 Table 4-2)
KS	0.664 in^3 (ACI SP-4 Table 4-3)
E	1,500,000 lbf/in^2 (ACI SP-4 Table 4-2)
Ib/Q	8.882 in^2 (ACI SP-4 Table 4-3)

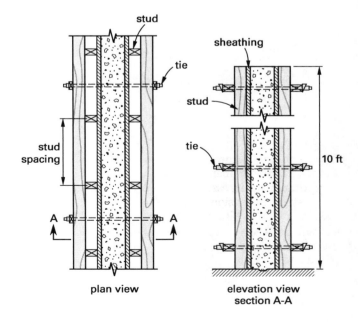

plan view   elevation view section A-A

Approximately what stud spacing should be used?

(A) 9 in

(B) 12 in

(C) 15 in

(D) 18 in

Hint: Consult ACI SP-4 and ASCE 37.

79. A concrete exterior wall is shown. The concrete is poured at a rate of 6 ft/hr at a temperature of 45°F. The spacing of the wales is 18 in, and the form tie spacing is 24 in. Form tie yield stress is 24 kips/in^2 with a strength reduction factor of 0.9. What is the minimum allowable form tie diameter?

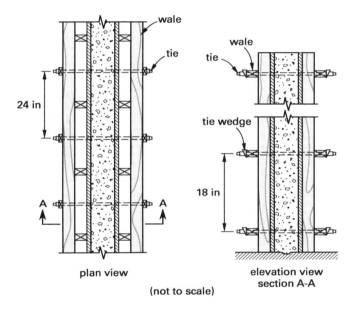

plan view

elevation view
section A-A

(not to scale)

(A) $^3/_8$ in

(B) $^1/_2$ in

(C) $^5/_8$ in

(D) $^3/_4$ in

Hint: The load a form tie can support is the product of yield stress and cross-sectional area.

80. A concrete wall will be poured under windy conditions. The wall formwork shown is braced using nominal 2 in × 4 in wooden members. The wooden members have a maximum compressive stress of 800 lbf/in^2, a maximum stress in tension of 750 lbf/in^2, and a modulus of elasticity of 1,200,000 lbf/in^2. The wind load at the top of the form is 80 lbf/ft, and the maximum slenderness ratio for the brace is 50. What is most nearly the maximum allowable spacing for the lateral braces?

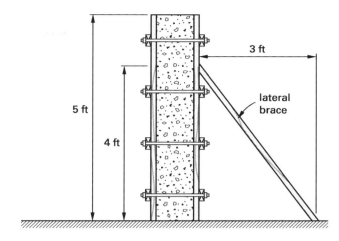

(A) 7.0 ft

(B) 10 ft

(C) 15 ft

(D) 25 ft

Hint: Find the lateral compressive load on the strut and the compressive load capacity of the brace.

81. According to ACI 347, which of the following statements about reshores is FALSE?

(A) They are vertical or inclined support members designed to carry the weight of formwork loads immediately above.

(B) They must be placed snugly under a stripped concrete slab.

(C) They are installed only after a concrete slab is able to support its own weight.

(D) They are placed after shores are removed.

Hint: Consult ACI 347.

82. Adding fly ash to a cement mixture will increase all of the following EXCEPT

(A) binding

(B) strength

(C) durability

(D) permeability

Hint: Review fly ash characteristics.

83. A wide flange steel beam (W24 × 68, unbraced span length 16 ft) is used to resist flexural loads. Using LRFD, the maximum moment on the beam is 600 ft-kips, and the yield strength is 50 kips/in². Does the beam have adequate strength to resist the load?

(A) Yes, the beam is adequate to carry 600 ft-kips.

(B) No, the beam needs to be braced to reduce the span length to 13.5 ft.

(C) No, the beam needs to be braced to reduce the span length to 11.5 ft.

(D) No, the beam needs to be braced to reduce the span length to 9.5 ft.

Hint: Consult Part 3 in the *AISC Manual.*

84. A sign is mounted onto a steel tube as shown. The tube is welded to a steel plate and anchored into a concrete base using anchor bolts installed on 10 in on center. The wind load is 40 lbf/ft². Using a factor of safety of 3, what is most nearly the required tensile capacity of each steel bolt?

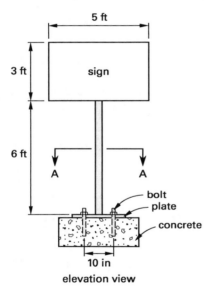

elevation view

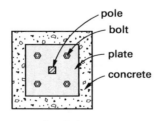

section A-A

(A) 2700 lbf

(B) 6500 lbf

(C) 8100 lbf

(D) 16,000 lbf

Hint: The bolts form a resisting moment, which counters the moment generated from the horizontal wind load.

85. A cofferdam with soldier piles spaced 6 ft apart is constructed in a river that has the following properties.

water surface	8 ft above riverbed
water flow	uniform depth, perpendicular to sheetpiles
water velocity, v	3 ft/sec
water density, ρ	62.4 lbm/ft³
drag coefficient, C_D	1.0

The maximum moment at each soldier pile that is created by a dynamic drag force is most nearly

(A) 1700 ft-lbf

(B) 2000 ft-lbf

(C) 5300 ft-lbf

(D) 16,000 ft-lbf

Hint: The water will form a dynamic force that will be applied in the mid-depth due to uniform flow.

86. A square spread footing for a column is constructed as shown. ($f'_c = 4000$ lbf/in², and $f_y = 60,000$ lbf/in².) Reinforcement is used to resist tension, whereas temperature and shrinkage steel is used in the compression side. The distance between the center of the bottom reinforcements and the bottom of the concrete is 3 in. What is most nearly the minimum area of tension reinforcement each way?

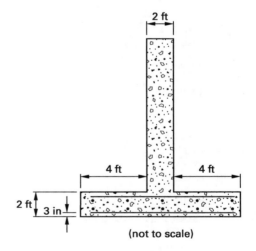

(not to scale)

(A) 8.0 in²

(B) 8.5 in²

(C) 9.5 in²

(D) 17 in²

Hint: Consult ACI 318 Chap. 10.

Worker Health, Safety, and Environment

87. Which statement is INCORRECT according to OSHA scaffolding regulations?

(A) Guardrails must be provided on scaffolding that is more than 6 ft above a lower level.

(B) Suspension ropes on scaffolds must be able to support at least six times the maximum intended load.

(C) The scaffold platform cannot deflect more than 1/60 of the span when loaded.

(D) The work area for each scaffold platform must be at least 18 in.

Hint: Consult OSHA Std. 1926.451.

88. According to OSHA excavation regulations, what is the maximum allowable slope for an excavation of 18 ft in sand?

(A) 90°

(B) 53°

(C) 45°

(D) 34°

Hint: Consult OSHA Std. 1926.652 App. B.

89. Which statement is INCORRECT according to OSHA safety management regulations?

(A) An employer must use OSHA forms 300, 300A, or 301 to record workplace injuries and illnesses.

(B) An employer does not have to keep safety records if it has fewer than 10 employees at all times during a calendar year.

(C) An employer must record motor vehicle accidents of employees commuting to or from work.

(D) An employer must orally report the death of an employee from a work-related accident to an OSHA office within eight hours of the accident.

Hint: Consult OSHA Std. 1904.

90. An OSHA compliance safety and health officer conducted an inspection of a workplace and found a violation. After some time, the OSHA officer returned to the workplace and found that the violation was not corrected for. The owner of the workplace will most likely be cited for which type of violation?

(A) repeated

(B) willful

(C) serious

(D) failure to abate

Hint: Consult OSHA Std. 3000-08R.

91. A company has an experience modification rate of 0.89. In comparison to similar businesses, this company has

(A) a worse-than-average record of claims due to work-related injuries

(B) a better-than-average record of claims due to work-related injuries

(C) been with the same insurer for a longer-than-average time

(D) been with the same insurer for a shorter-than-average time

Hint: The experience modification rate (EMR) is a factor used by insurance companies when determining insurance premiums.

92. A large construction company has 800 full-time employees who work 50 weeks per year. Last year, the company reported a total of 12 minor injuries and no severe injuries, job-related illnesses, or fatalities. The injury and illness incidence rate for the company last year was most nearly

(A) 0.015 injuries/yr

(B) 1.4 injuries/yr

(C) 1.5 injuries/yr

(D) 12 injuries/yr

Hint: The injury and illness incidence rate is a scaled number that represents the rate of injuries and illnesses per year per 100 employees.

Other Topics

93. Which of the following dewatering options are commonly used during groundwater exclusion?

I.   sump pumps

II. slurry walls

III. steel sheet-piles

IV. well points

V.   freeze walls

VI. ejector wells

(A) I, II, and III

(B) II, III, and V

(C) I, IV, and VI

(D) II, III, and VI

Hint: Groundwater exclusion is not the same as groundwater pumping.

94. A rectangular drilling pattern is used for rock blasting. The explosives will be placed in drilled holes with diameters of 3 in that are spaced every 8 ft. The holes are drilled to 25 ft deep. If the blast will produce a satisfactory rock break at 22 ft deep, the rock volume yielded per foot of drilling will be most nearly

(A) 0.52 yd³/ft

(B) 1.6 yd³/ft

(C) 2.1 yd³/ft

(D) 2.4 yd³/ft

Hint: Calculate the rock volume produced by blasting and divide by the length of hole drilled.

95. A mechanically stabilized retaining wall with metal strips extending into the soil is shown. The lateral soil loads are resisted by the friction between the soil and the metal strips. The strips are located at the centers of the face units. The specific weight of the soil is 120 lbf/ft³, and the friction angle of the soil is 30°. The face units are 3 ft × 3 ft, and the metal strips are 2 ft wide horizontally. If the required safety factor is 3, what is most nearly the minimum required length of the shaded metal strip?

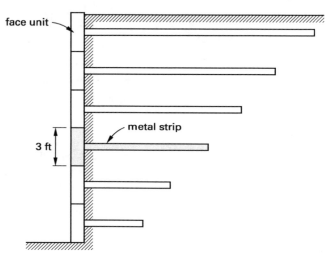

(A) 3.9 ft

(B) 4.3 ft

(C) 8.2 ft

(D) 12 ft

Hint: Obtain the active horizontal soil pressure, and then determine the length of strip needed for the factor of safety.

96. A deep foundation is needed for a high-rise building project. The provided empirical formula (which is not dimensionally consistent) will be used to estimate the safe driving load of the concrete piles.

$$R = \left(\frac{2E}{S+0.1}\right)\left(\frac{W_r + KW_p}{W_r + W_p}\right)$$

The variables are defined as

R = safe driving load (lbf)
E = energy of hammer (ft-lbf)
S = average penetration per blow, last six blows (in)
W_r = weight of hammer ram (lbf)
W_p = weight of pile, including driving appurtenances (lbf)
K = coefficient of restitution, from the table provided

weight of pile per foot, including driving appurtenances (lbf/ft)	coefficient of restitution, K
≤ 50	0.2
50 to 100	0.4
> 100	0.6

The concrete piles used are 8 in square and 50 ft long, and the specific weight of the concrete is 150 lbf/ft³. The contractor was provided with the hydraulic pile driver information as given.

pile driver energy = 15,000 ft-lbf
ram weight = 3500 lbf
weight of driving appurtenances = 1000 lbf
average penetration from last six blows = 0.25 in/blow

The safe driving load of the piles is most nearly

(A) 57 kips

(B) 61 kips

(C) 67 kips

(D) 69 kips

Hint: Use the formula provided.

97. The water tower shown is located on the roof of an apartment complex in a region that has a ground snow load of 50 lbf/ft². What is most nearly the drift height in the leeward direction?

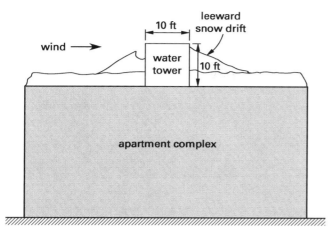

elevation

(A) 1.4 ft

(B) 1.8 ft

(C) 2.5 ft

(D) 10 ft

Hint: Consult ASCE 7 Chap. 7.

98. A simply supported steel beam bears both a uniform load and a concentrated load. The beam is W12 × 26 with a length of 12 ft. The unfactored uniform dead load is 0.8 kips/ft. The unfactored concentrated live load, which is located at the midpoint of the beam, is 5 kips. The total deflection tolerance is $L/360$, and deflection occurs at midspan. Using LRFD, which statement is true?

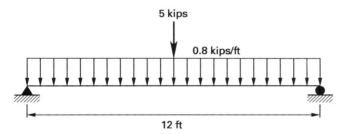

(A) The maximum deflection is 0.118 in, which is smaller than allowable deflection of 0.4 in, so the design is OK.

(B) The maximum deflection is 0.160 in, which is smaller than allowable deflection of 0.4 in, so the design is OK.

(C) The maximum deflection is 0.118 in, which is greater than allowable deflection of 0.033 in, so a larger section is needed.

(D) The maximum deflection is 0.160 in, which is greater than allowable deflection of 0.033 in, so a larger section is needed.

Hint: Consult *AISC Manual* Part 3 and ASCE 7 Chap. 2.

99. A cross-sectional area of a concrete beam is shown. Three no. 6 bars with 2 in of concrete clear cover are used for reinforcement. The yield strength is 4 kips/in^2 for the concrete and 60 kips/in^2 for the steel. Using LRFD, what is most nearly the nominal moment capacity of the concrete beam?

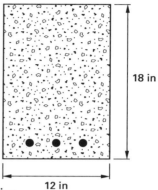

(A) 87 ft-kips

(B) 97 ft-kips

(C) 100 ft-kips

(D) 110 ft-kips

Hint: Consult ACI 318 Chap. 10.

100. A traffic lane must be closed to allow for construction on a street with a 40 mph speed limit. What is most nearly the required minimum merging taper length?

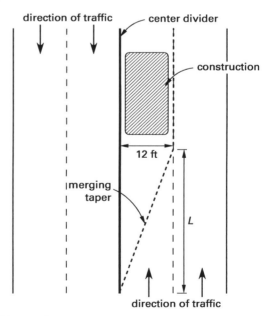

(A) 100 ft

(B) 160 ft

(C) 320 ft

(D) 480 ft

Hint: Consult MUTCD Chap. 6.

Solutions
Breadth Problems

1. The excavation area is an irregular shape. To calculate the volume of an irregular shape, the entire area must first be divided into sections. Then, the volume for each section is calculated. Connect point A and point D to divide the entire area into two trapezoids.

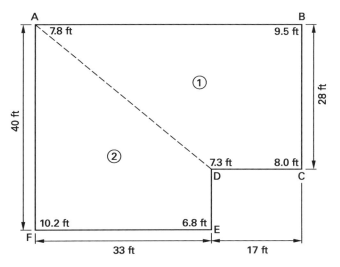

Calculate the volume of trapezoid 1. The average excavation depth, d, of trapezoid 1 is

$$d_{\text{trap }1} = \frac{A + B + C + D}{4}$$
$$= \frac{7.8 \text{ ft} + 9.5 \text{ ft} + 8.0 \text{ ft} + 7.3 \text{ ft}}{4}$$
$$= 8.15 \text{ ft}$$

The area, A, of trapezoid 1 is

$$A_{\text{trap }1} = \left(\frac{a + b}{2}\right)h = \left(\frac{17 \text{ ft} + (17 \text{ ft} + 33 \text{ ft})}{2}\right)(28 \text{ ft})$$
$$= 938 \text{ ft}^2$$

Therefore, the volume, V, of trapezoid 1 is

$$V_{\text{trap }1} = d_{\text{trap }1}A_{\text{trap }1} = (8.15 \text{ ft})(938 \text{ ft}^2)$$
$$= 7644.7 \text{ ft}^3$$

Calculate the volume of trapezoid 2. The average excavation depth of trapezoid 2 is

$$d_{\text{trap }2} = \frac{A + D + E + F}{4}$$
$$= \frac{7.8 \text{ ft} + 7.3 \text{ ft} + 6.8 \text{ ft} + 10.2 \text{ ft}}{4}$$
$$= 8.03 \text{ ft}$$

The area of trapezoid 2 is

$$A_{\text{trap }2} = \left(\frac{a + b}{2}\right)h = \left(\frac{12 \text{ ft} + 40 \text{ ft}}{2}\right)(33 \text{ ft})$$
$$= 858 \text{ ft}^2$$

Therefore, the volume of trapezoid 2 is

$$V_{\text{trap }2} = d_{\text{trap }2}A_{\text{trap }2} = (8.03 \text{ ft})(858 \text{ ft}^2)$$
$$= 6889.7 \text{ ft}^3$$

The total bank cubic yards, BCY, is

$$\text{BCY} = V_{\text{trap }1} + V_{\text{trap }2} = \frac{7644.7 \text{ ft}^3 + 6889.7 \text{ ft}^3}{27 \ \frac{\text{ft}^3}{\text{yd}^3}}$$
$$= 538.31 \text{ yd}^3$$

Calculate the loose cubic yards, LCY, with the 8% soil swell factor, SF.

$$\text{LCY} = (1 + \text{SF})\text{BCY} = (1 + 0.08)(538.31 \text{ yd}^3)$$
$$= 581.38 \text{ yd}^3$$

The truck's capacity is 12 yd³. Find the total number of loads needed to transport the excavated soil.

$$\text{number of loads} = \frac{581.38 \text{ yd}^3}{12 \ \frac{\text{yd}^3}{\text{load}}}$$
$$= 48.45 \text{ loads} \quad (50 \text{ loads})$$

The answer is (C).

Why Other Options Are Wrong

(A) This incorrect option used the shallowest depth of each trapezoid instead of the average depth. Additionally, it did not account for the swell factor.

(B) This incorrect option did not account for the swell factor.

(D) This incorrect option used the greatest depth of each trapezoid instead of the average depth.

2. The cut volume, V_{cut}, from sta 2+00 to sta 2+50 is

$$V_{cut} = \left(\frac{C_1 + C_2}{2}\right)(50 \text{ ft}) = \frac{\left(\dfrac{170 \text{ ft}^2 + 230 \text{ ft}^2}{2}\right)(50 \text{ ft})}{27 \dfrac{\text{ft}^3}{\text{yd}^3}}$$

$$= 370.4 \text{ yd}^3$$

Find the compacted soil volume, CCY, for the fill from sta 2+00 to sta 2+50.

$$CCY_{fill} = \left(\frac{F_1 + F_2}{2}\right)(50 \text{ ft}) = \frac{\left(\dfrac{200 \text{ ft}^2 + 120 \text{ ft}^2}{2}\right)(50 \text{ ft})}{27 \dfrac{\text{ft}^3}{\text{yd}^3}}$$

$$= 296.3 \text{ yd}^3$$

Compute the bank volume, BCY, needed for fill using a shrinkage factor, DF, of 12%.

$$BCY = \frac{CCY}{1 - DF} = \frac{296.3 \text{ yd}^3}{1 - 0.12}$$

$$= 336.7 \text{ yd}^3$$

Therefore, the net cut volume is

$$V_{net\ cut} = V_{cut} - BCY = 370.4 \text{ yd}^3 - 336.7 \text{ yd}^3$$

$$= 33.7 \text{ yd}^3 \quad (35 \text{ yd}^3)$$

The answer is (A).

Why Other Options Are Wrong

(B) This incorrect option did not account for the shrinkage factor.

(C) This incorrect option calculated the fill volume.

(D) This incorrect option calculated the cut volume.

3. The compacted soil volume, CCY, for the building's foundation is

$$CCY = lwd = \frac{(200 \text{ ft})(100 \text{ ft})(5 \text{ ft})}{27 \dfrac{\text{ft}^3}{\text{yd}^3}}$$

$$= 3703.7 \text{ yd}^3$$

The soil shrinkage factor, DF, is 10%. The bank volume, BCY, of the soil is then

$$BCY = \frac{CCY}{1 - DF} = \frac{3703.7 \text{ yd}^3}{1 - 0.10}$$

$$= 4115.2 \text{ yd}^3$$

The soil swell factor, SF, is 12%. The loose volume of the soil, LCY, is then

$$LCY = (1 + SF)BCY = (1 + 0.12)(4115.2 \text{ yd}^3)$$

$$= 4609.1 \text{ yd}^3 \quad (4610 \text{ yd}^3)$$

The soil density, ρ, is 120 lbm/ft^3. Therefore, the soil weight, W_{soil}, is

$$W_{soil} = \frac{(LCY)\rho g}{g_c}$$

$$= \frac{(4609.1 \text{ yd}^3)\left(120 \dfrac{\text{lbm}}{\text{ft}^3}\right)\left(27 \dfrac{\text{ft}^3}{\text{yd}^3}\right)\left(32.2 \dfrac{\text{ft}}{\text{sec}^2}\right)}{32.2 \dfrac{\text{lbm-ft}}{\text{lbf-sec}^2}}$$

$$= 14{,}933{,}333 \text{ lbf}$$

The water content of the soil is 7%, which means the ratio of the water weight to the soil weight is 7/100. Therefore, the water weight, W_{water}, is

$$W_{water} = (14{,}933{,}333 \text{ lbf})\left(\frac{0.07}{0.07 + 1.0}\right)$$

$$= 976{,}947 \text{ lbf} \quad (977{,}000 \text{ lbf})$$

The answer is (C).

Why Other Options Are Wrong

(A) This incorrect option did not include the swell factor.

(B) This incorrect option did not include the swell factor, and the water weight was calculated by merely multiplying the soil's total weight by 7%.

(D) This incorrect option calculated the water weight by merely multiplying the soil's total weight by 7%.

4. The weight of the borrow soil that will fill the truck to its capacity of 12 yd^3 is

$$W_{\text{soil}} = V_{\text{soil}}\gamma_{\text{soil}} = \frac{(12 \text{ yd}^3)\left(115 \frac{\text{lbf}}{\text{ft}^3}\right)\left(27 \frac{\text{ft}^3}{\text{yd}^3}\right)}{2000 \frac{\text{lbf}}{\text{ton}}}$$

$$= 18.63 \text{ tons}$$

The truck's capacity is 15 tons, so it will not be able to carry 12 yd^3 of the borrow soil. The maximum volume that the truck can carry is

$$V_{\text{truck}} = \frac{W_{\text{truck}}}{\gamma_{\text{soil}}} = \frac{\left(15 \frac{\text{tons}}{\text{load}}\right)\left(2000 \frac{\text{lbf}}{\text{ton}}\right)}{\left(115 \frac{\text{lbf}}{\text{ft}^3}\right)\left(27 \frac{\text{ft}^3}{\text{yd}^3}\right)}$$

$$= 9.67 \text{ yd}^3/\text{load}$$

The bank volume, BCY, from the borrow pit is

$$\text{BCY} = lwd = \frac{(200 \text{ ft})(150 \text{ ft})(3 \text{ ft})}{27 \frac{\text{ft}^3}{\text{yd}^3}}$$

$$= 3333.33 \text{ yd}^3$$

The soil swell factor, SF, is 12%. The loose volume, LCY, of excavated soil is

$$\text{LCY} = (1 + \text{SF})\text{BCY} = (1 + 0.12)(3333.33 \text{ yd}^3)$$

$$= 3733.33 \text{ yd}^3$$

Therefore, the number of loads the truck will need to make is

$$\frac{\text{LCY}}{V_{\text{truck}}} = \frac{3733.33 \text{ yd}^3}{9.67 \frac{\text{yd}^3}{\text{load}}}$$

$$= 386.1 \text{ loads} \quad (390 \text{ loads})$$

The answer is (D).

Why Other Options Are Wrong.

(A) This incorrect option did not account for the truck's maximum capacity of 15 tons.

(B) This incorrect option did not include the soil swell factor in its calculations.

(C) This incorrect option incorrectly rounded the number of required loads.

5. Locate the chords from points PC and PT to the midpoint of the arc.

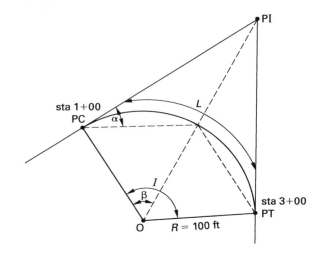

Find the length of the arc, L, from PC to PT.

$$L = \text{PT} - \text{PC} = (\text{sta } 3{+}00) - (\text{sta } 1{+}00)$$

$$= 200 \text{ ft}$$

Find the intersection angle, I. R is the radius of the curve, which is given as 100 ft.

$$L = \frac{2\pi RI}{360°}$$

$$I = \frac{L(360°)}{2\pi R} = \frac{(200 \text{ ft})(360°)}{2\pi(100 \text{ ft})}$$

$$= 114.6°$$

The deflection angle, α, is the angle between the tangent and the chord on either side of the curve. The deflection angle α is half of angle β.

Angle β is half of the angle that the arc encloses.

$$\beta = \frac{114.6°}{2} = 57.3°$$

Therefore, the deflection angle is

$$\alpha = \frac{\beta}{2} = \frac{57.3°}{2}$$

$$= 28.65° \quad (30°)$$

The answer is (A).

Why Other Options Are Wrong

(B) This incorrect option used half of the intersection angle between the two tangents instead of using half of the deflection angle.

(C) This incorrect option calculated half of the arc angle instead of calculating half of the deflection angle.

(D) This incorrect option calculated the angle of the entire arc instead of calculating the deflection angle.

6. The elevation difference, y, is related to the gradient change, A, the horizontal distance, x, and the curve length, L.

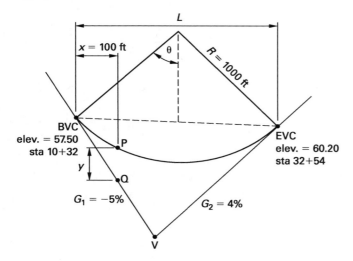

Find the gradient change.

$$A = G_2 - G_1 = 4\% - (-5\%)$$
$$= 9\%$$

Calculate the length of the curve between BVC and EVC.

$$L = \text{EVC} - \text{BVC} = (\text{sta } 32{+}54) - (\text{sta } 10{+}32)$$
$$= \text{sta } 22{+}22 \quad (2222 \text{ ft})$$

The elevation difference, y, between the road and the tangent is

$$y = \frac{Ax^2}{200L} = \frac{(9)(100 \text{ ft})^2}{(200)(2222 \text{ ft})}$$
$$= 0.203 \text{ ft}$$

The elevation at point Q can be calculated by the slope and the BVC station.

$$\text{elev}_Q = \text{elev}_{\text{BVC}} + G_1 x = 57.50 \text{ ft} + \frac{(-5\%)(100 \text{ ft})}{100\%}$$
$$= 52.50 \text{ ft}$$

The elevation of point P is

$$\text{elev}_P = \text{elev}_Q + y = 52.50 \text{ ft} + 0.203 \text{ ft}$$
$$= 52.70 \text{ ft}$$

The answer is (D).

Alternative Solution

Find the rate of change.

$$R = \frac{G_2 - G_1}{L} = \frac{4\% - (-5\%)}{22.22 \text{ sta}}$$
$$= 0.405 \text{ \%/sta} \quad [\text{same as } 0.405 \text{ ft/sta}^2]$$

One station equals 100 ft, and a grade of -5% equals -5 ft/sta. Therefore, the elevation at point P is

$$\text{elev}_P = \frac{Rx^2}{2} + G_1 x + \text{elev}_{\text{BVC}}$$
$$= \frac{\left(0.405 \dfrac{\text{ft}}{\text{sta}^2}\right)(1 \text{ sta})^2}{2}$$
$$+ \left(-5 \dfrac{\text{ft}}{\text{sta}}\right)(1 \text{ sta}) + 57.50 \text{ ft}$$
$$= 52.70 \text{ ft}$$

The answer is (D).

Why Other Options Are Wrong

(A) This incorrect option subtracted the difference of the curve and the tangent from the elevation of point Q instead of adding it.

(B) This incorrect option assumed the change of slope was 1% (5% − 4%) instead of 9% (5% − (−4%)).

(C) This incorrect option assumed the change of slope was 5% (i.e., the initial slope) instead of 9%.

7. The cross-sectional area of the concrete is

$$A = A_{\text{stem}} + A_{\text{foot}} = (2 \text{ ft})(4 \text{ ft}) + (4 \text{ ft})(2 \text{ ft})$$
$$= 16 \text{ ft}^2$$

The volume of concrete needed per foot of wall length is

$$V_{\text{per foot}} = A\left(1 \dfrac{\text{ft}}{\text{ft}}\right) = (16 \text{ ft}^2)\left(1 \dfrac{\text{ft}}{\text{ft}}\right)$$
$$= 16 \text{ ft}^3/\text{ft}$$

The length perimeter of the building is

$$P = 40 \text{ ft} + 50 \text{ ft} + 50 \text{ ft} + \frac{\pi(40 \text{ ft})}{2}$$

$$= 202.8 \text{ ft}$$

The amount of concrete needed for the building's perimeter is

$$V = V_{\text{per foot}} P = \frac{\left(16 \frac{\text{ft}^3}{\text{ft}}\right)(202.8 \text{ ft})}{27 \frac{\text{ft}^3}{\text{yd}^3}}$$

$$= 120.2 \text{ yd}^3 \quad (120 \text{ yd}^3)$$

The answer is (B).

Why Other Options Are Wrong

(A) This incorrect option used a diameter of 20 ft instead of 40 ft when calculating the length of the semicircular arc.

(C) This incorrect option used the total height (6 ft) as the wall stem height.

(D) This incorrect option used the entire circumference of the circle as the length of the semicircular arc.

8. Calculate the volume of formwork needed per foot of foundation length. For the two sides of the footing,

$$V_{\text{footing}} = 2hLt = (2)(4 \text{ ft})(1 \text{ ft})\left(\frac{2 \text{ in}}{12 \frac{\text{in}}{\text{ft}}}\right)$$

$$= 1.33 \text{ ft}^3/\text{ft}$$

For the two sides of the stem,

$$V_{\text{stem}} = 2hLt = (2)(8 \text{ ft})(1 \text{ ft})\left(\frac{2 \text{ in}}{12 \frac{\text{in}}{\text{ft}}}\right)$$

$$= 2.67 \text{ ft}^3/\text{ft}$$

The total volume per foot of foundation length is

$$V_{\text{per foot}} = V_{\text{footing}} + V_{\text{stem}} = 1.33 \frac{\text{ft}^3}{\text{ft}} + 2.67 \frac{\text{ft}^3}{\text{ft}}$$

$$= 4 \text{ ft}^3/\text{ft}$$

Convert to units of board foot measure. A board foot is not a unit of length; rather, it is used with lumber to measure volume. There are 144 in^3 in 1 bd-ft.

$$V_{\text{per foot}} = \frac{\left(4 \frac{\text{ft}^3}{\text{ft}}\right)\left(12 \frac{\text{in}}{\text{ft}}\right)^3}{144 \frac{\text{in}^3}{\text{bd-ft}}}$$

$$= 48 \text{ bd-ft/ft}$$

The length of the building perimeter is

$$P = 80 \text{ ft} + 100 \text{ ft} + 100 \text{ ft} + \frac{\pi(80 \text{ ft})}{2}$$

$$= 405.66 \text{ ft}$$

Therefore, the amount of formwork needed is

$$V_{\text{total}} = V_{\text{per foot}} P = \left(48 \frac{\text{bd-ft}}{\text{ft}}\right)(405.66 \text{ ft})$$

$$= 19,471.86 \text{ bd-ft} \quad (19,500 \text{ bd-ft})$$

The answer is (B).

Why Other Options Are Wrong

(A) This incorrect option calculated the perimeter of the building by using the diameter of the circle for the length of the semicircular arc.

(C) This incorrect option calculated the perimeter of the building by using the entire circumference of the circle for the length of the semicircular arc.

(D) This incorrect option calculated the volume of the stem by using the total height (12 ft) as the stem height.

9. The number of project members and the hourly wage for each member is known. Therefore, the labor hour, LH, can be found by averaging the hourly costs for all project members.

$$LH = \frac{\begin{array}{l}(1 \text{ foreman})\left(50 \frac{\$}{\text{hr}}\right) + (2 \text{ carpenters})\left(45 \frac{\$}{\text{hr}}\right) \\ + (1 \text{ painter})\left(35 \frac{\$}{\text{hr}}\right) + (1 \text{ laborer})\left(25 \frac{\$}{\text{hr}}\right)\end{array}}{5 \text{ members}}$$

$$= \$40/\text{hr}$$

The overtime labor hour, LH$_{\text{overtime}}$, for the entire project team is then

$$LH_{\text{overtime}} = (1.50)\left(40 \frac{\$}{\text{hr}}\right)$$

$$= \$60/\text{hr}$$

The answer is (B).

Why Other Options Are Wrong

(A) This incorrect option calculated an average using only one project member per category.

(C) This incorrect option summed the hourly rates in each category instead of solving for the labor hour.

(D) This incorrect option calculated the total cost of the crew instead of the labor hour.

10. To find the weekly cost, first calculate the real hourly rate for the welder. The real hourly rate is the sum of the welder's base hourly rate ($30/hr) and the hourly cost of all related insurance premiums.

Social security is 6% of the welder's base hourly rate. The hourly cost is

$$\left(30 \; \frac{\$}{\text{hr}}\right)(0.06) = \$1.80/\text{hr}$$

Workers' compensation is 5% of the welder's base hourly rate. The hourly cost is

$$\left(30 \; \frac{\$}{\text{hr}}\right)(0.05) = \$1.50/\text{hr}$$

Liability insurance is 4% of the welder's base hourly rate. The hourly cost is

$$\left(30 \; \frac{\$}{\text{hr}}\right)(0.04) = \$1.20/\text{hr}$$

Therefore, the base hourly rate is

$$30 \; \frac{\$}{\text{hr}} + 1.80 \; \frac{\$}{\text{hr}} + 1.50 \; \frac{\$}{\text{hr}} + 1.20 \; \frac{\$}{\text{hr}} = \$34.50/\text{hr}$$

Calculate the total weekly cost for a 40 hour week.

$$\left(34.50 \; \frac{\$}{\text{hr}}\right)\left(40 \; \frac{\text{hr}}{\text{wk}}\right) = \$1380/\text{wk}$$

The company will pay 75% of the welder's health and dental insurance.

$$\frac{\left(400 \; \dfrac{\$}{\text{mo}}\right)(0.75)\left(7 \; \dfrac{\text{days}}{\text{wk}}\right)}{30 \; \dfrac{\text{days}}{\text{mo}}} = \$70/\text{wk}$$

Therefore, the total weekly cost to the company is

$$1380 \; \frac{\$}{\text{wk}} + 70 \; \frac{\$}{\text{wk}} = \$1450/\text{wk} \quad (\$1500/\text{wk})$$

The answer is (C).

Why Other Options Are Wrong

(A) This incorrect option multiplied the base rate by 40 hours without accounting for other weekly costs.

(B) This incorrect option did not add the cost of health and dental insurance to the total cost.

(D) This incorrect option did not convert the monthly cost of health and dental insurance to the weekly cost before adding it to the total cost.

11. Construction sequencing follows the logic that one activity cannot start until the previous activity has commenced or finished. For this problem, excavation (activity III) must first be performed before the foundation can be poured. After the foundation is poured (activity V), columns can be constructed to support the slabs (activity IV). Next, the forms and shoring of slabs will be built, and the slab will be poured (activity I). For the steel roof, joints and composite metal roofing are set up (activity VI). Once this is done, the roof concrete can be poured directly onto the metal roofing (activity II). This will create a composite reaction through shear studs that engages the concrete strength with the corrugated steel sheets and steel joists.

The answer is (D).

Why Other Options Are Wrong

(A) This incorrect option presented an incorrect sequence.

(B) This incorrect option presented an incorrect sequence.

(C) This incorrect option presented an incorrect sequence.

12. The sequence of activities for constructing a single-span bridge is

1. The piles are driven down to the designated depth (activity VI).

2. Formwork is prepared for the pouring of concrete pile caps that will support the abutments of the bridge (activity VIII).

3. The abutment forms are set up with reinforcement in place, and the concrete is poured and cured for the abutments (activity IV).

4. Steel girders or stringers are placed on bearing pads. They extend from abutment to abutment in a single-span bridge, and floor beams or diaphragms are placed perpendicular to girders at a set distance for structural integrity and stability (activity II).

5. The metal form is placed on top of steel girders with shear studs to engage composite reaction (activity VII).

6. Steel reinforcement is placed on the metal form before the concrete deck is placed (activity V).

7. Concrete is poured on top of the form to the desired depth (activity III).

8. When the concrete is cured, layers of pavement are placed on top of the deck for protection (activity I).

The answer is (A).

Why Other Options Are Wrong

(B) This incorrect option presented an incorrect sequence.

(C) This incorrect option presented an incorrect sequence.

(D) This incorrect option presented an incorrect sequence.

13. Draw trial vertical lines through the shading in the bar chart. The trial line that cuts the most shaded lines is week 6, which has six activities in progress at the same time (wood framing, sheathing and drywall, plumbing and electrical, windows and doors, carpentry, and kitchen). Therefore, the owner should visit the construction site during week 6.

The answer is (B).

Why Other Options Are Wrong

(A) This incorrect option has only five activities in progress (fewer than the six activities in week 6).

(C) This incorrect option has only five activities in progress (fewer than the six activities in week 6).

(D) This incorrect option has only five activities in progress (fewer than the six activities in week 6).

14. To find the total float of each activity, the earliest start (ES), earliest finish (EF), latest start (LS), and latest finish (LF) must be determined.

Determine the ES and EF. Calculate the earliest possible time that an activity can start after the previous activity finishes. The EF time will be the ES time plus the duration, D, of the activity. If an activity has more than one predecessor, the ES time must be after the activity finishing latest. That is,

$$ES = EF_{predecessor}$$
$$EF = ES + D$$

Tabulate the ES and EF dates for each activity.

activity	earliest start (day)	earliest finish (day)
AB	0	4
BC	4	12
BD	4	7
CD	12	17
DE	17	21
CF	12	18
EF	21	30
EG	21	27
FG	30	37

Create a chart with time (in days) as the horizontal axis and activity as the vertical axis.

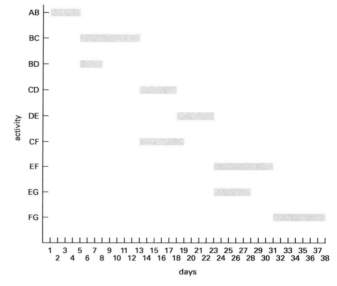

Using the table provided in the problem, calculate the total number of resources for each week.

Week 1 includes activities AB, BC, and BD. The total number of resources is $7 + 5 + 16 = 28$.

Week 2 includes activities BC, CD, and CF. The total number of resources is $5 + 11 + 7 = 23$.

Week 3 includes activities CD, DE, and CF. The total number of resources is $11 + 6 + 7 = 24$.

Week 4 includes activities EF and EG. The total number of resources is $6 + 9 = 15$.

Week 5 includes activities EF and FG. The total number of resources is $6 + 5 = 11$.

Week 6 includes only activity FG. The total number of resources is 5.

Therefore, the most resources are used during week 1.

The answer is (A).

Why Other Options Are Wrong

(B) This incorrect option chose week 1, which uses only 23 resources.

(C) This incorrect option chose week 3, which uses only 24 resources.

(D) This incorrect option chose week 4, which uses only 15 resources.

15. *Earned value management* (EVM) is used to assess a project's performance by comparing the work performed by the work planned. In EVM, the *cost performance index* (CPI) represents the relationship between the actual cost expended and the earned value. The CPI value is obtained by dividing the budgeted cost of worked performed (BCWP) by the actual cost of work performed (ACWP). A CPI greater than or equal to 1.0 suggests positive cash flow. The *schedule variance* (SV) is the difference between the value of work performed for a given period and the value of the work as planned. It is measured in value (e.g., dollars), not time. The *budgeted cost of work scheduled* (BCWS) is a project's spending plan as a function of performance and schedule.

The *schedule performance index* (SPI) measures schedule effectiveness. SPI is calculated by dividing the BCWP by the BCWS. An SPI greater than (not less than) 1.0 indicates a project is ahead of schedule.

The answer is (B).

Why Other Options Are Wrong

(A) This incorrect option assumed a CPI greater than or equal to 1.0 does not suggest profit.

(C) This incorrect option assumed the SV is measured in time, not dollars.

(D) This incorrect option assumed the BCWS is not a spending plan for a project.

16. The illustration shown represents the general behavior of a project's *total cost*, which is calculated by summing its direct and indirect costs. The *optimum* project is one that is conducted during *normal time* (i.e., the amount of time a project would normally take to complete), as it has the optimal working pace and is conducted with the minimum cost. Once a project passes the optimum point, labor and equipment costs will increase.

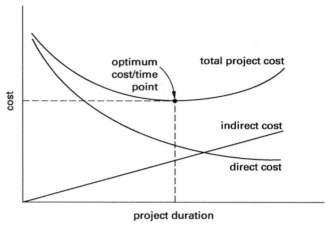

The construction schedule is usually planned according to normal time. If the project is completed before the optimum point, then the project is completed ahead of schedule. Usually, completing a project ahead of schedule requires increasing a project's productivity by increasing its resources and decreasing its duration. The shortest amount of time a project can be completed is called *crash time*. Under crash time, a project's original critical path will be shortened, but in most cases, the sequence of its activities will not change.

The answer is (B).

Why Other Options Are Wrong

(A) This incorrect option is a true statement about time-cost trade-off analysis.

(C) This incorrect option is a true statement about time-cost trade-off analysis.

(D) This incorrect option is a true statement about time-cost trade-off analysis.

17. Convert sieve size to grain size (in millimeters).

$$
\begin{aligned}
\text{no. } 4 &= 4.75 \text{ mm} \\
\text{no. } 10 &= 2.00 \text{ mm} \\
\text{no. } 40 &= 0.425 \text{ mm} \\
\text{no. } 100 &= 0.15 \text{ mm} \\
\text{no. } 200 &= 0.075 \text{ mm}
\end{aligned}
$$

Plot the sieve analysis on a grain size distribution graph.

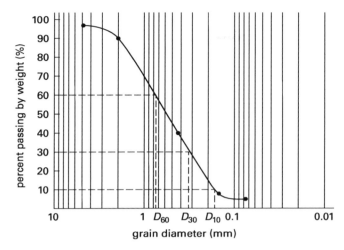

Use the USCS soil classification table to determine whether the soil sample is coarse-grained. (See *Table for Sol. 17.*)

From the table, the soil is a coarse-grained sample because only 5% of it passed the no. 200 sieve. (This also means the plastic limit and the liquid limit are not important when classifying this sample.)

The no. 4 sieve is used to identify whether a coarse soil sample is sand or gravel. 97% of the soil sample (more than 50%) passed the no. 4 sieve, so the sample is classified as sand.

Only 5% of the soil passed the no. 200 sieve. Use the USCS table to classify whether the soil can be SW or SP. To distinguish SP from SW, the coefficient of uniformity and the coefficient of curvature must be calculated.

Find the coefficient of uniformity, C_u. D_{60} is the grain diameter (in millimeters) corresponding to 60% passing, and D_{10} is the grain diameter (in millimeters) corresponding to 10% passing. From the graph, $D_{60} = 0.71$ mm, and $D_{10} = 0.18$ mm.

$$C_u = \frac{D_{60}}{D_{10}} = \frac{0.71 \text{ mm}}{0.18 \text{ mm}} = 3.9$$

Table for Sol. 17

major division		group symbol	finer than 200 sieve (%)	supplementary requirements	soil description
			laboratory classification criteria		
coarse-grained (over 50% by weight coarser than no. 200 sieve)	gravelly soils (over half of coarse fraction larger than no. 4)	GW	0–5[*]	D_{60}/D_{10} greater than 4 $D_{30}^2 \% (D_{60}D_{10})$ between 1 and 3	well-graded gravels, sandy gravels
		GP	0–5[*]	not meeting above gradation requirements for GW	gap-graded or uniform gravels, sandy gravels
		GM	12 or more[*]	PI less than 4 or below A-line	silty gravels, silty sandy gravels
		GC	12 or more[*]	PI over 7 and above A-line	clayey gravels, clayey sandy gravels
	sandy soils (over half of coarse fraction finer than no. 4)	SW	0–5[*]	D_{60}/D_{10} greater than 6 $D_{30}^2 \% (D_{60}D_{10})$ between 1 and 3	well-graded, gravelly sands
		SP	0–5[*]	not meeting above gradation requirements for SW	gap-graded or uniform sands, gravelly sands
		SM	12 or more[*]	PI less than 4 or below A-line	silty sand
		SC	12 or more[*]	PI over 7 and above A-line	clayey sands, clayey gravelly sands
fine-grained (over 50% by weight finer than no. 200 sieve)	low compressibility (liquid limit less than 50)	ML	plasticity chart		silts, very fine sands, silty or clayey fine sands, micaceous silts
		CL	plasticity chart		low plasticity clays, sandy or silty clays
		OL	plasticity chart, organic odor or color		organic silts and clays of high plasticity
	high compressibility (liquid limit 50 or more)	MH	plasticity chart		micaceous silts, diatomaceous silts, volcanic ash
		CH	plasticity chart		highly plastic clays and sandy clays
		OH	plasticity chart, organic odor or color		organic silts and clays of high plasticity
soils with fibrous organic matter		Pt	fibrous organic matter; will char, burn, or glow		peat, sandy peats, and clayey peats

[*]For soils having 5–12% passing the no. 200 sieve, use a dual symbol such as GW-GC.

Find the coefficient of curvature, C_c. D_{30} is the grain diameter (in millimeters) corresponding to 30% passing. From the graph, $D_{30} = 0.34$ mm.

$$C_c = \frac{D_{30}^2}{D_{10}D_{60}} = \frac{(0.34 \text{ mm})^2}{(0.18 \text{ mm})(0.71 \text{ mm})}$$
$$= 0.91$$

From the table, a soil with a coefficient of curvature between 1 and 3 is considered well-graded as long as the coefficient of uniformity is greater than 4 for gravels and greater than 6 for sands. In this case, $C_u < 1$ and $C_c < 6$, so the soil sample is poorly graded. Therefore, the soil classification is SP.

The answer is (A).

Why Other Options Are Wrong

(B) This incorrect option assumed the soil was not poorly graded.

(C) This incorrect option assumed the soil met the "sand with fines" criterion.

(D) This incorrect option assumed the soil met the "sands with fines" criterion.

18. A *concrete slump test* (ASTM C143) measures the water content in unhardened concrete to determine the consistency between batches. Conducting the test requires

- an even surface (e.g., plywood)

- a slump cone, which is a 12 in tall hollow metal truncated cone that is open at both ends, with a bottom diameter of 8 in and a top diameter of 4 in

- a tamping rod that is $\frac{5}{8}$ in diameter and 24 in long

- a measuring device (often a tape measure) at least 12 in long with markings to at least $\frac{1}{4}$ in

Concrete is poured into the hollow cone in three layers of approximately one-third increments. Each layer is tamped 25 times with a round, spherical-nosed steel rod that is $\frac{5}{8}$ in diameter and 24 in long. The excess concrete on top is removed after the last tamping. Then, the metal cone is lifted vertically without rotation. The cone should be removed cautiously so that concrete is not sheared. The height of the concrete is measured, and the difference between the height of the concrete and 12 in is noted. This difference, the *slump*, provides a measure of the consistency and workability of the concrete.

For valid results, the entire concrete slump test must be completed within $2\frac{1}{2}$ minutes. Therefore, option C is not part of the test.

The answer is (C).

Why Other Options Are Wrong

(A) This incorrect option is a step of the concrete slump test.

(B) This incorrect option is a step of the concrete slump test.

(D) This incorrect option is a step of the concrete slump test.

19. The actual weight of the worker and the weight of the equipment are inconsequential to the individual personnel load. ASCE 37 Sec. 4.1.1 provides load definitions for construction and defines the *individual personnel load* as "a concentrated load of 250 lbf that includes the weight of one person plus equipment carried by that person or equipment that can be readily picked up by a single person without assistance." That is, each worker is assumed to be 200 lbf with 50 lbf of equipment weight, or 250 lbf.

The answer is (C).

Why Other Options Are Wrong

(A) This incorrect option subtracted the equipment's weight from the worker's weight.

(B) This incorrect option added the equipment's weight to the worker's weight.

(D) This incorrect option added the equipment's weight to the ASCE 37 defined individual personnel load.

20. ASCE 37 Sec. 4.4 requires a horizontal construction load be applied to temporary or partially complete structures. The horizontal construction load is found as the greater of the following criteria.

- for wheeled vehicles transporting materials, 20% for a single vehicle or 10% for two or more vehicles of the fully loaded vehicle weight

- equipment reactions as defined in ASCE 37 Sec. 4.6

- 50 lbf per person applied at the level of the platform in any direction

- 2% of the total vertical load

Earthquake loads are not considered when calculating the horizontal construction load.

The answer is (D).

Why Other Options Are Wrong

(A) This incorrect option assumed total vertical loads are not considered in horizontal construction load calculations.

(B) This incorrect option assumed personnel loads are not considered in horizontal construction load calculations.

(C) This incorrect option assumed equipment reactions are not considered in horizontal construction calculations.

Solutions
Depth Problems

21. In a mass diagram, the volume on the y-axis is cumulative. (This is different from a direct profile diagram, in which elevations of a cross section are shown.)

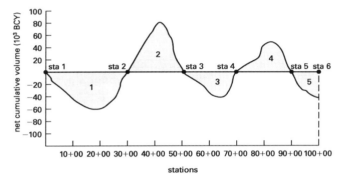

In region 1, the mass diagram slopes down and comes back up to zero. This means that this region needs fill at the beginning, in the portion where the cumulative value goes down. The curve reaches a low value of -60×10^3 BCY. Then cut occurs, which produces excess soil and results in an increase of cumulative volume. The cut volume finally breaks even to bring the curve back to zero at sta 30+00. Therefore, the volume of cut for the section from sta 0+00 to sta 30+00 is 60×10^3 BCY.

In region 2, from sta 30+00 to sta 50+00, the mass diagram curves up and then slopes back down to zero. This means that the cut section does not end at sta 30+00, but continues in region 2 until the point where the curve reaches a high value of 80×10^3 BCY and begins decreasing, at which point another fill occurs. The cumulative volume then goes down until the volumes of cut and fill break even to zero. Therefore, the cut volume in region 2 is 80×10^3 BCY.

Using similar calculations, the cut volumes in region 3 and region 4 are 40×10^3 BCY and 50×10^3 BCY, respectively. In region 5, only fill occurs since the curve slopes down for the entire region.

The total cut volume is then

$$V_{\text{total}} = V_{\text{region 1}} + V_{\text{region 2}} + V_{\text{region 3}} + V_{\text{region 4}}$$
$$= (60 \times 10^3 \text{ BCY}) + (80 \times 10^3 \text{ BCY})$$
$$\quad + (40 \times 10^3 \text{ BCY}) + (50 \times 10^3 \text{ BCY})$$
$$= 230 \times 10^3 \text{ BCY} \quad (2.3 \times 10^5 \text{ BCY})$$

The answer is (C).

Why Other Options Are Wrong

(A) This incorrect option included only the sag area of the mass diagram as the cut volume.

(B) This incorrect option included only the crest area of the mass diagram as the cut volume.

(D) This incorrect option included all the maximum and minimum points.

22. The bank volume, BCY, of the excavated soil is

$$\text{BCY} = \frac{(350 \text{ ft})(4 \text{ ft})(3 \text{ ft})}{27 \, \dfrac{\text{ft}^3}{\text{yd}^3}}$$
$$= 155.6 \text{ yd}^3$$

The loose bank volume, LCY, with a soil swell factor, SF, of 18% is

$$\text{LCY} = \text{BCY}\left(1 + \frac{\text{SF}}{100\%}\right) = (155.6 \text{ yd}^3)\left(1 + \frac{18\%}{100\%}\right)$$
$$= 183.6 \text{ yd}^3$$

Find the base, B, of the triangular spoil bank using the following formula. L is the length of the bank, and ϕ is the angle of repose.

$$B = \sqrt{\frac{4(\text{LCY})}{L \tan \phi}} = \sqrt{\frac{(4)(183.6 \text{ yd}^3)\left(27 \, \dfrac{\text{ft}^3}{\text{yd}^3}\right)}{(80 \text{ ft}) \tan 35°}}$$
$$= 18.81 \text{ ft}$$

Therefore, the height, h, of the triangular bank is

$$h = \frac{B \tan \phi}{2} = \frac{(18.81 \text{ ft}) \tan 35°}{2}$$
$$= 6.59 \text{ ft} \quad (7 \text{ ft})$$

The answer is (D).

Why Other Options Are Wrong

(A) This incorrect option did not include the swell factor in its calculations; furthermore, it calculated the trench cross-sectional area as a triangle instead of a rectangle.

(B) This incorrect option calculated the trench cross-sectional area as a triangle instead of a rectangle.

(C) This incorrect option calculated the trench cross-sectional area correctly as a rectangle, but it did not include the swell factor in its calculations.

23. (a) Find the dry density of the borrow soil, $\rho_{d,\text{borrow}}$.

$$\rho_{d,\text{borrow}} = \frac{(\text{SG})\rho_w}{1 + \frac{w(\text{SG})}{S}} = \frac{(2.75)\left(62.4\ \frac{\text{lbm}}{\text{ft}^3}\right)}{1 + \frac{(0.15)(2.75)}{0.70}}$$
$$= 108.0\ \text{lbm/ft}^3$$

After compaction, the soil's volume is 1000 yd^3 with a bulk density of 115 lbm/ft^3 and a water content of 18%. Therefore, the total mass, m, of the soil (including the solids and water) is

$$m = V_{\text{soil}}\rho_{\text{compaction}} = (1000\ \text{yd}^3)\left(115\ \frac{\text{lbm}}{\text{ft}^3}\right)\left(27\ \frac{\text{ft}^3}{\text{yd}^3}\right)$$
$$= 3{,}105{,}000\ \text{lbm}$$

The optimal water content, w_{optimal}, is 18%. Therefore, the compacted soil's water mass, $m_{w,\text{optimal}}$, is

$$m_{w,\text{optimal}} = \frac{m w_{\text{optimal}}}{1 + w_{\text{optimal}}} = \frac{(3{,}105{,}000\ \text{lbm})(0.18)}{1 + 0.18}$$
$$= 473{,}644\ \text{lbm}$$

The mass of solids in the soil is

$$m_s = m - m_{w,\text{optimal}} = 3{,}105{,}000\ \text{lbm} - 473{,}644\ \text{lbm}$$
$$= 2{,}631{,}356\ \text{lbm}$$

The volume of soil, V_{soil}, needed from the borrow pit is

$$V_{\text{soil}} = \frac{m}{\rho_{\text{borrow}}} = \frac{2{,}631{,}356\ \text{lbm}}{\left(108.0\ \frac{\text{lbm}}{\text{ft}^3}\right)\left(27\ \frac{\text{ft}^3}{\text{yd}^3}\right)}$$
$$= 902.39\ \text{yd}^3 \quad (900\ \text{yd}^3)$$

(b) The borrow soil's water mass, $m_{w,\text{borrow}}$, is

$$m_{w,\text{borrow}} = m_s w_{\text{borrow}} = (2{,}631{,}356\ \text{lbm})(0.15)$$
$$= 394{,}703.4\ \text{lbm}$$

Calculate the number of gallons of water that need to be added to the compacted soil.

$$m_{w,\text{optimal}} - m_{w,\text{borrow}} = \frac{473{,}644\ \text{lbm} - 394{,}703.4\ \text{lbm}}{8.34\ \dfrac{\text{lbm}}{\text{gal}}}$$
$$= 9465.3\ \text{gal} \quad (9500\ \text{gal})$$

The answer is (A).

Why Other Options are Wrong

(B) This incorrect option did not convert the borrow volume using the density of the borrow pit.

(C) This incorrect option calculated the borrow volume by dividing the total mass by the dry density, and the volume of water was incorrectly calculated by assuming water content was the weight of water over the weight of solids.

(D) This incorrect option calculated the borrow volume by dividing the total mass by the dry density, not bulk density.

24. The loose volume, $V_{\text{loose,borrow}}$, of a conical spoil pile is obtained by using the formula

$$V_{\text{loose,borrow}} = \tfrac{1}{3}\pi\left(\frac{D}{2}\right)^2\left(\frac{D\tan\phi}{2}\right) = \tfrac{1}{24}\pi D^3 \tan\phi$$
$$= \tfrac{1}{24}\pi(5\ \text{ft})^3 \tan 40°$$
$$= 13.73\ \text{ft}^3$$

Find the required volume of compacted soil, $V_{\text{compact,req}}$.

$$V_{\text{compact,req}} = \frac{bhl}{2} = \frac{(3\ \text{ft})(2\ \text{ft})(4\ \text{ft})}{2}$$
$$= 12\ \text{ft}^3$$

Find the bank volume, $V_{\text{bank,borrow}}$, of the borrow soil. The swell factor, SF, is given as 15%.

$$V_{\text{bank,borrow}} = \frac{\text{LCY}}{1 + \text{SF}} = \frac{13.73\ \text{ft}^3}{1 + 0.15}$$
$$= 11.94\ \text{ft}^3$$

Find the compacted volume of the borrow soil, $V_{\text{compact,borrow}}$. The shrinkage factor, DF, is given as 10%.

$$V_{\text{compact,borrow}} = (1 - \text{DF})\text{BCY} = (1 - 0.10)(11.94\ \text{ft}^3)$$
$$= 10.75\ \text{ft}^3$$

Since the required volume of soil is 12 ft^3, additional soil will need to be purchased. The volume of additional compacted soil is

$$V_{\text{compact,req}} - V_{\text{compact,borrow}} = 12\ \text{ft}^3 - 10.75\ \text{ft}^3$$
$$= 1.25\ \text{ft}^3$$

Therefore, there is 1.25 ft^3 of soil that needs to be purchased.

The answer is (C).

Why Other Options Are Wrong

(A) This incorrect option did not account for the shrinkage factor.

(B) This incorrect option did not account for the swell and shrinkage factors.

(D) This incorrect option calculated the compacted soil as a rectangle instead of a triangle.

25. Find the elevation difference between the backsight (BS) of point A and the foresight (FS) of point B.

$$\text{elev}_{AB} = \text{BS}_A - \text{FS}_B = 5.12 \text{ ft} - 3.85 \text{ ft}$$
$$= 1.27 \text{ ft} \quad [\text{point B is higher}]$$

Find the elevation difference between the backsight of point B and the foresight of point C.

$$\text{elev}_{BC} = \text{BS}_B - \text{FS}_C = 2.52 \text{ ft} - 6.72 \text{ ft}$$
$$= -4.20 \text{ ft} \quad [\text{point C is lower}]$$

Therefore, the elevation difference between point A and point C is

$$\text{elev}_{AC} = \text{elev}_{AB} + \text{elev}_{BC} = 1.27 \text{ ft} + (-4.20 \text{ ft})$$
$$= -2.93 \text{ ft}$$

The elevation at point A is 105 ft. Therefore, the elevation at point C is

$$\text{elev}_C = \text{elev}_A + \text{elev}_{AC} = 105 \text{ ft} + (-2.93 \text{ ft})$$
$$= 102.07 \text{ ft} \quad (102 \text{ ft})$$

The answer is (B).

Why Other Options Are Wrong

(A) This incorrect option used -1.27 ft instead of 1.27 ft as the elevation difference between points A and B.

(C) This incorrect option added 2.93 ft instead of -2.93 ft to the elevation at point A.

(D) This incorrect option used 4.20 ft instead of -4.20 ft as the elevation difference between points B and C.

26. Slope stakes indicate points where designed cut and fill slopes intersect existing ground. Slope stakes are marked with either "C" for cut or "F" for fill. On the slope stake provided, the first marking reads C-7.5/15, which means that at 15 ft from the slope stake, the ground surface will be cut 7.5 ft below the elevation at the slope stake.

Find the cut area from the slope stake to a distance of 15 ft by multiplying the horizontal distance, d, by the average cut depth, h.

$$A_{\text{cut},1} = d\left(\frac{h_1 + h_2}{2}\right) = (15 \text{ ft})\left(\frac{0 \text{ ft} + 7.5 \text{ ft}}{2}\right)$$
$$= 56.25 \text{ ft}^2$$

The second marking on the slope stake reads C-10/25, which means that at 25 ft from the slope stake, the ground surface will be cut 10 ft below the elevation at the slope stake. Find the second cut area.

$$A_{\text{cut},2} = (10 \text{ ft})\left(\frac{7.5 \text{ ft} + 10 \text{ ft}}{2}\right)$$
$$= 87.5 \text{ ft}^2$$

The third marking on the slope stake reads C-5/35, which means that at 35 ft from the slope stake, the ground surface will be cut 5 ft below the elevation at the slope stake. Find the third cut area.

$$A_{\text{cut},3} = (10 \text{ ft})\left(\frac{10 \text{ ft} + 5 \text{ ft}}{2}\right)$$
$$= 75 \text{ ft}^2$$

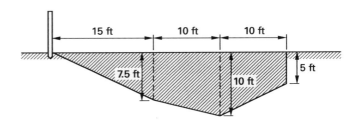

Therefore, the total cross-sectional area that will be excavated is

$$A_{\text{cut,total}} = A_{\text{cut},1} + A_{\text{cut},2} + A_{\text{cut},3}$$
$$= 56.25 \text{ ft}^2 + 87.5 \text{ ft}^2 + 75 \text{ ft}^2$$
$$= 218.75 \text{ ft}^2 \quad (220 \text{ ft}^2)$$

The answer is (A).

Why Other Options Are Wrong

(B) This incorrect option used the elevation at each marking as the reference for the next cut area. The elevation of the slope stake should be used as the reference elevation for all cut areas.

(C) This incorrect option interpreted each horizontal distance referenced from the slope stake as the full length of a cut section. The horizontal distances marked by the slope stake should be measured from the slope stake, not from the previous marking.

(D) This incorrect option combined the errors made in option B and option C.

27. From the mass diagram, the distance between the balance points is three stations. The average haul distance is half of the distance between balance points, or 1.5 stations.

Draw a line that represents 1.5 stations on the mass diagram. The cumulative volume where the line intercepts the mass diagram is the overhaul volume, or 500 yd^3.

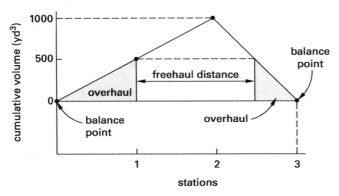

Determine the contractor's freehaul and overhaul costs using the overhaul volume of 500 yd^3.

$$\text{freehaul} = (500 \text{ yd}^3)\left(2 \ \frac{\$}{\text{yd}^3}\right)$$
$$= \$1000$$

$$\text{overhaul} = (500 \text{ yd}^3)\left(3 \ \frac{\$}{\text{yd}^3}\right)$$
$$= \$1500$$

The contractor's total haul cost is $\$1000 + \$1500 = \$2500$.

Determine the subcontractor's haul cost.

$$C_{\text{subcontractor}} = (1000 \text{ yd}^3)\left(2 \ \frac{\$}{\text{yd}^3}\right)$$
$$= \$2000$$

Therefore, the cost difference between the contractor's haul cost and the subcontractor's haul cost is $\$2500 - \$2000 = \$500$. The subcontractor's haul cost is $\$500$ less than the contractor's haul cost.

The answer is (B).

Why Other Options Are Wrong

(A) This incorrect option did not include the contractor's freehaul cost.

(C) This incorrect option did not include the subcontractor's freehaul cost.

(D) This incorrect option calculated the contractor's total haul costs.

28. Section C_1 represents a cut in the shape of a semicircle with a radius, r, of 40 ft. The volume, V, of C_1 is

$$V_{C1} = \tfrac{1}{2}(\pi r^2)w_{\text{road}} = \frac{\left(\frac{1}{2}\right)\pi(40 \text{ ft})^2(12 \text{ ft})}{27 \ \frac{\text{ft}^3}{\text{yd}^3}}$$
$$= 1117.01 \text{ yd}^3$$

Section C_2 represents a cut in the shape of a quarter-circle with a radius of 60 ft. The volume of C_2 is

$$V_{C2} = \tfrac{1}{4}(\pi r^2)w_{\text{road}} = \frac{\left(\frac{1}{4}\right)\pi(60 \text{ ft})^2(12 \text{ ft})}{27 \ \frac{\text{ft}^3}{\text{yd}^3}}$$
$$= 1256.63 \text{ yd}^3$$

The total cut volume is

$$V_{C,\text{total}} = V_{C1} + V_{C2} = 1117.01 \text{ yd}^3 + 1256.63 \text{ yd}^3$$
$$= 2373.64 \text{ yd}^3$$

Section F_1 represents a fill area in the shape of a semicircle with a radius of 20 ft. The volume of F_1 is

$$V_{F1} = \tfrac{1}{2}(\pi r^2)w_{\text{road}} = \frac{\left(\frac{1}{2}\right)\pi(20 \text{ ft})^2(12 \text{ ft})}{27 \ \frac{\text{ft}^3}{\text{yd}^3}}$$
$$= 279.25 \text{ yd}^3$$

Section F_2 represents a fill area in the shape of a semicircle with a radius of 30 ft. The volume of F_2 is

$$V_{F2} = \tfrac{1}{2}(\pi r^2)w_{\text{road}} = \frac{\left(\frac{1}{2}\right)\pi(30 \text{ ft})^2(12 \text{ ft})}{27 \ \frac{\text{ft}^3}{\text{yd}^3}}$$
$$= 628.32 \text{ yd}^3$$

Therefore, the total fill volume is

$$V_{F,\text{total}} = V_{F1} + V_{F2} = 279.25 \text{ yd}^3 + 628.32 \text{ yd}^3$$
$$= 907.57 \text{ yd}^3$$

The soil's excess bank volume, BCY, is

$$BCY = V_{C,total} - V_{F,total} = 2373.64 \text{ yd}^3 - 907.25 \text{ yd}^3$$
$$= 1466.39 \text{ yd}^3$$

The swell factor, SF, is given as 12%. So, the soil's excess loose volume, LCY, is

$$LCY = (1 + SF)BCY = (1 + 0.12)(1466.39 \text{ yd}^3)$$
$$= 1642.36 \text{ yd}^3$$

Therefore, the number of truck loads, x, that will be required to transport the soil is

$$x = \frac{LCY}{\text{truck capacity}} = \frac{1642.36 \text{ yd}^3}{10 \frac{\text{yd}^3}{\text{load}}}$$
$$= 164.23 \text{ loads} \quad (160 \text{ loads})$$

The answer is (C).

Why Other Options Are Wrong

(A) This incorrect option considered only the fill volumes.

(B) This incorrect option did not include the swell factor in the calculation.

(D) This incorrect option considered only the cut volumes.

29. The steel joists are W12 × 26, so each joist weighs 26 lbf/ft. Each joist is 20 ft long, and there are eight joists. Therefore, the weight, W, of the steel joists is

$$W_{\text{joists}} = \left(26 \frac{\text{lbf}}{\text{ft}}\right)\left(20 \frac{\text{ft}}{\text{joist}}\right)(8 \text{ joists})$$
$$= 4160 \text{ lbf}$$

The steel beams are W14 × 30, so each beam weighs 30 lbf/ft. Each beam is 20 ft long, and there are 14 beams. Therefore, the weight of the steel beams is

$$W_{\text{beams}} = \left(30 \frac{\text{lbf}}{\text{ft}}\right)\left(20 \frac{\text{ft}}{\text{beam}}\right)(14 \text{ beams})$$
$$= 8400 \text{ lbf}$$

The steel columns are W12 × 24, so each column weighs 24 lbf/ft. Each column is 12 ft long, and there are 12 columns. Therefore, the weight of the steel columns is

$$W_{\text{columns}} = \left(24 \frac{\text{lbf}}{\text{ft}}\right)\left(12 \frac{\text{ft}}{\text{column}}\right)(12 \text{ columns})$$
$$= 3456 \text{ lbf}$$

Calculate the total weight for the building's frame.

$$W_{\text{total}} = W_{\text{joists}} + W_{\text{beams}} + W_{\text{columns}}$$
$$= \frac{4160 \text{ lbf} + 8400 \text{ lbf} + 3456 \text{ lbf}}{2000 \frac{\text{lbf}}{\text{ton}}}$$
$$= 8.01 \text{ tons} \quad (8.0 \text{ tons})$$

The answer is (C).

Why Other Options Are Wrong

(A) This incorrect option only included one level of the steel building.

(B) This incorrect option omitted the middle beam and so used only 12 beams to calculate the total beam weight.

(D) This incorrect option calculated the length of the columns as 20 ft instead of as 12 ft.

30. To find the total surface area that will be covered with bricks, calculate the area of the house's perimeter walls and subtract the area of the doors and windows. The surface area, A, of each long wall is

$$A_{\text{long wall}} = Lh = (70 \text{ ft})(15 \text{ ft})$$
$$= 1050 \text{ ft}^2$$

In measuring the surface area of the short walls, the width of one brick at each end of the wall can be disregarded, as this area will be provided by the bricks in the long walls. Therefore, the surface area of each short wall is

$$A_{\text{short wall}} = (L - 2w_{\text{brick}})h$$
$$= \left(40 \text{ ft} - (2)\left(\frac{3.75 \text{ in}}{12 \frac{\text{in}}{\text{ft}}}\right)\right)(15 \text{ ft})$$
$$= 590.63 \text{ ft}^2$$

The surface area of each door is

$$A_{\text{door}} = hw = (8 \text{ ft})(3 \text{ ft})$$
$$= 24 \text{ ft}^2$$

The surface area of each window is

$$A_{\text{window}} = hw = (4 \text{ ft})(3 \text{ ft})$$
$$= 12 \text{ ft}^2$$

Therefore, the total wall area is

$$A_{total} = 2A_{long\,wall} + 2A_{short\,wall} - 2A_{door} - 12A_{window}$$

$$= \left(\begin{array}{c} (2)(1050 \text{ ft}^2) + (2)(590.63 \text{ ft}^2) \\ - (2)(24 \text{ ft}^2) - (12)(12 \text{ ft}^2) \end{array} \right) \left(12 \, \frac{\text{in}}{\text{ft}}\right)^2$$

$$= 444{,}853.44 \text{ in}^2$$

The thickness of the mortar between two bricks is $^1/_2$ in, or $^1/_4$ in for each brick. The surface area of one brick, including a $^1/_4$ in thickness of mortar all around, is

$$A_{brick} = (L_{brick} + 2t_{mortar})(h_{brick} + 2t_{mortar})$$

$$= \big(8 \text{ in} + (2)(0.25 \text{ in})\big)\big(2.25 \text{ in} + (2)(0.25 \text{ in})\big)$$

$$= 23.38 \text{ in}$$

Therefore, the number of bricks, n, that is needed to construct the four walls is

$$n_{bricks} = \frac{A_{total}}{A_{brick}} = \frac{444{,}853.44 \text{ in}^2}{23.38 \text{ in}^2}$$

$$= 19{,}027.09 \quad (19{,}000 \text{ bricks})$$

The answer is (B).

Why Other Options Are Wrong

(A) This incorrect option calculated the thickness of mortar as $^1/_2$ in around each brick instead of $^1/_4$ in.

(C) This incorrect option did not subtract the area for the door and window openings.

(D) The incorrect option did not account for the thickness of mortar.

31. Find the total number of bricks required. Calculate the gross surface area, A, of each wall, as if there were no openings.

$$A_{gross\,wall} = wh = (15 \text{ ft})(12 \text{ ft})$$

$$= 180 \text{ ft}^2$$

The surface area of each window is

$$A_{window} = wh = (3 \text{ ft})(4 \text{ ft})$$

$$= 12 \text{ ft}^2$$

The surface area of each mechanical conduit opening is

$$A_{mech} = \frac{\pi d^2}{4} = \frac{\pi (1 \text{ ft})^2}{4}$$

$$= 0.785 \text{ ft}^2$$

The surface area of each louver is

$$A_{louver} = wh = (1.5 \text{ ft})(1.5 \text{ ft})$$

$$= 2.25 \text{ ft}^2$$

The net surface area of each wall is then

$$A_{net\,wall} = A_{gross\,wall} - 2A_{window} - 4A_{mech} - A_{louver}$$

$$= 180 \text{ ft}^2 - (2)(12 \text{ ft}^2) - (4)(0.785 \text{ ft}^2) - 2.25 \text{ ft}^2$$

$$= 150.61 \text{ ft}^2$$

The total surface area of all eight walls is

$$A_{total} = 8A_{net\,wall} = (8)(150.61 \text{ ft}^2)$$

$$= 1205 \text{ ft}^2$$

The thickness of the mortar between two bricks is $^1/_2$ in, so the area of the mortar will be $^1/_4$ in around each brick. The surface area of one brick with mortar is

$$A_{brick/mortar} = (L + 2t_{mortar})(h + 2t_{mortar})$$

$$= \frac{\begin{array}{c}\big(7.625 \text{ in} + (2)(0.25 \text{ in})\big) \\ \times \big(2.25 \text{ in} + (2)(0.25 \text{ in})\big)\end{array}}{\left(12 \, \frac{\text{in}}{\text{ft}}\right)^2}$$

$$= 0.155 \text{ ft}^2$$

The number of bricks, n, needed to construct the four walls is

$$n_{bricks} = \frac{A_{total}}{A_{brick}} = \frac{1205 \text{ ft}^2}{0.155 \text{ ft}^2}$$

$$= 7774$$

The surface area of one brick without mortar is

$$A_{brick} = Lh = \frac{(7.625 \text{ in})(2.25 \text{ in})}{\left(12 \, \frac{\text{in}}{\text{ft}}\right)^2}$$

$$= 0.119 \text{ ft}^2$$

The volume of mortar used per brick is

$$V_{per\,brick} = (A_{brick/mortar} - A_{brick})t_{wall}$$

$$= (0.155 \text{ ft}^2 - 0.119 \text{ ft}^2)\left(\frac{3.625 \text{ in}}{12 \, \frac{\text{in}}{\text{ft}}}\right)$$

$$= 0.0109 \text{ ft}^3$$

Therefore, the total amount of mortar needed is

$$V_{\text{total}} = n_{\text{bricks}} V_{\text{per brick}} = \dfrac{(7774)(0.0109 \text{ ft}^3)}{27 \, \dfrac{\text{ft}^3}{\text{yd}^3}}$$

$$= 3.13 \text{ yd}^3 \quad (3.0 \text{ yd}^3)$$

The answer is (B).

Why Other Options Are Wrong

(A) This incorrect option calculated the number of bricks for only one wall.

(C) This incorrect option included only one window, one mechanical opening, and one louver per wall.

(D) This incorrect option did not account for any openings.

32. With one vertical no. 6 reinforcement bar at each end of the wall and the rest spaced throughout the wall at 12 in intervals, the number of vertical bars needed for a 100 ft wall is 101. (The number of spaces between bars is one less than the number of bars—two bars will contain one space, three bars will contain two spaces, and so on. If the wall is 100 ft long and each space is 1 ft, there is room for 100 spaces between bars, so there will be 101 bars.)

The length of each vertical reinforcement bar is

$$6 \text{ ft} + 2 \text{ ft} + 2 \text{ ft} = 10 \text{ ft}$$

The weight per linear foot is obtained by multiplying the volume per linear foot by the unit weight of steel. For no. 6 bars,

$$w_{\text{no.6}} = V\gamma = AL\gamma = \left(\dfrac{\pi d^2}{4}\right) L\gamma$$

$$= \left(\dfrac{\pi \left(\dfrac{0.75 \text{ in}}{12 \, \dfrac{\text{in}}{\text{ft}}}\right)^2}{4}\right) (1 \text{ ft})\left(490 \, \dfrac{\text{lbf}}{\text{ft}^3}\right)$$

$$= 1.503 \text{ lbf/ft}$$

The total weight of the vertical reinforcement is

$$W_{\text{no.6}} = n_{\text{bars}} L w_{\text{no.6}} = (101 \text{ bars})\left(10 \, \dfrac{\text{ft}}{\text{bar}}\right)\left(1.503 \, \dfrac{\text{lbf}}{\text{ft}}\right)$$

$$= 1518 \text{ lbf}$$

From the stem height shown in the illustration, seven horizontal no. 5 reinforcement bars are needed in the stem, and 18 longitudinal no. 5 reinforcement bars are

needed in the footing, for a total of 25 no. 5 bars. Each bar is 100 ft long. For no. 5 bars, the weight per linear foot is

$$w_{\text{no.5}} = \left(\dfrac{\pi d^2}{4}\right) L\gamma = \left(\dfrac{\pi \left(\dfrac{0.625 \text{ in}}{12 \, \dfrac{\text{in}}{\text{ft}}}\right)^2}{4}\right) (1 \text{ ft})\left(490 \, \dfrac{\text{lbf}}{\text{ft}^3}\right)$$

$$= 1.044 \text{ lbf/ft}$$

The total weight of the horizontal and longitudinal reinforcement bars is

$$W_{\text{no.5}} = n_{\text{bars}} L w_{\text{no.5}} = (25 \text{ bars})\left(100 \, \dfrac{\text{ft}}{\text{bar}}\right)\left(1.044 \, \dfrac{\text{lbf}}{\text{ft}}\right)$$

$$= 2610 \text{ lbf}$$

With two transverse no. 7 reinforcement bars in the footing, the number of transverse bars needed for a 100 ft footing is 202. Each bar is 8 ft long. For no. 7 bars, the weight per linear foot is

$$w_{\text{no.7}} = \left(\dfrac{\pi d^2}{4}\right) L\gamma = \left(\dfrac{\pi \left(\dfrac{0.875 \text{ in}}{12 \, \dfrac{\text{in}}{\text{ft}}}\right)^2}{4}\right) (1 \text{ ft})\left(490 \, \dfrac{\text{lbf}}{\text{ft}^3}\right)$$

$$= 2.046 \text{ lbf/ft}$$

The total weight of the transverse reinforcement bars is

$$W_{\text{no.7}} = n_{\text{bars}} L w_{\text{no.7}} = (202 \text{ bars})\left(8 \, \dfrac{\text{ft}}{\text{bar}}\right)\left(2.046 \, \dfrac{\text{lbf}}{\text{ft}}\right)$$

$$= 3306 \text{ lbf}$$

Therefore, the total weight for the reinforcement in the retaining wall is

$$W_{\text{total}} = W_{\text{no.5}} + W_{\text{no.6}} + W_{\text{no.7}}$$

$$= 2610 \text{ lbf} + 1518 \text{ lbf} + 3306 \text{ lbf}$$

$$= 7434 \text{ lbf} \quad (7400 \text{ lbf})$$

The answer is (D).

Why Other Options Are Wrong

(A) This incorrect option calculated only half the required reinforcement in the footing (reinforcement for both the top and bottom is needed).

(B) This incorrect option used 6 ft as the length of the vertical reinforcement in the stem.

(C) This incorrect option did not include the hooked length of the vertical reinforcement in the stem.

33. In the United States, steel members are specified using their depth, d (in inches), and weight per unit length, w (in pounds-force per foot). The letter designation indicates the member type. For example, W18 × 50 represents a wide-flange steel beam that has a depth of approximately 18 in and has a unit load of 50 lbf/ft. If a member's length is known, the total weight, W, for each member can be determined.

Top chords

There are two W18 × 50 top chords per truss. The total weight of the top chords is

$$W_{\text{top}} = 2Lw = (2)(10 \text{ ft})\left(50 \ \frac{\text{lbf}}{\text{ft}}\right)$$
$$= 1000 \text{ lbf}$$

Therefore, the total load, P, from the top chords of both trusses is

$$P_{\text{top}} = 2W_{\text{top}} = (2)(1000 \text{ lbf})$$
$$= 2000 \text{ lbf}$$

Bottom chords

There are four W24 × 68 bottom chords per truss. The total weight of the bottom chords is

$$W_{\text{bottom}} = 4Lw = (4)(10 \text{ ft})\left(68 \ \frac{\text{lbf}}{\text{ft}}\right)$$
$$= 2720 \text{ lbf}$$

Therefore, the total load from the bottom chords of both trusses is

$$P_{\text{bottom}} = 2W_{\text{bottom}} = (2)(2720 \text{ lbf})$$
$$= 5440 \text{ lbf}$$

Vertical members

There are three C15 × 50 vertical members. The total weight for the vertical members is

$$W_{\text{vertical}} = 3Lw = (3)(10 \text{ ft})\left(50 \ \frac{\text{lbf}}{\text{ft}}\right)$$
$$= 1500 \text{ lbf}$$

Therefore, the total load from the vertical members of both trusses is

$$P_{\text{vertical}} = 2W_{\text{vertical}} = (2)(1500 \text{ lbf})$$
$$= 3000 \text{ lbf}$$

Diagonal members

The length, L, of the diagonal members is unknown, so use the formula for the length of a diagonal to find it.

$$L_{\text{diagonal}} = L_{\text{bottom chord}}\sqrt{2} = (10 \text{ ft})\sqrt{2}$$
$$= 14.14 \text{ ft}$$

There are four W21 × 48 diagonal members. The total weight of the diagonal members is

$$W_{\text{diagonal}} = 4Lw = (4)(14.14 \text{ ft})\left(48 \ \frac{\text{lbf}}{\text{ft}}\right)$$
$$= 2714.88 \text{ lbf}$$

The total load on the two trusses from the diagonal members is

$$P_{\text{diagonal}} = 2W_{\text{diagonal}} = (2)(2714.88 \text{ lbf})$$
$$= 5430 \text{ lbf}$$

Floor beams

There are five W24 × 104 floor beams. Therefore, the total weight of the floor beams is

$$W_{\text{floor}} = 5Lw = (5)(15 \text{ ft})\left(104 \ \frac{\text{lbf}}{\text{ft}}\right)$$
$$= 7800 \text{ lbf}$$

Find the weight per linear foot, w, of the bridge deck. The specific weight, γ, for normal weight concrete is 150 lbf/ft³. The thickness, t, of the bridge deck is given as 8 in.

$$w_{\text{concrete}} = Lt_{\text{deck}}\gamma_{\text{concrete}} = (15 \text{ ft})\left(\dfrac{8 \text{ in}}{12 \ \frac{\text{in}}{\text{ft}}}\right)\left(150 \ \frac{\text{lbf}}{\text{ft}^3}\right)$$
$$= 1500 \text{ lbf/ft}$$

Since the length of the bridge is 40 ft, the bridge deck's total load is

$$P_{\text{deck}} = L_{\text{deck}}w_{\text{concrete}} = (40 \text{ ft})\left(1500 \ \frac{\text{lbf}}{\text{ft}}\right)$$
$$= 60{,}000 \text{ lbf}$$

The total dead load, D, on both abutments is

$$D_{\text{total}} = P_{\text{top}} + P_{\text{bottom}} + P_{\text{vertical}} + P_{\text{diagonal}}$$
$$+ W_{\text{floor}} + P_{\text{deck}}$$
$$= 2000 \text{ lbf} + 5440 \text{ lbf} + 3000 \text{ lbf} + 5430 \text{ lbf}$$
$$+ 7800 \text{ lbf} + 60{,}000 \text{ lbf}$$
$$= 83{,}670 \text{ lbf}$$

However, the bridge is symmetrical. Therefore, each truss will take half the load. So, the dead load on one abutment is

$$D_{\text{per abutment}} = \frac{D_{\text{total}}}{2} = \frac{83,670 \text{ lbf}}{2}$$
$$= 41,835 \text{ lbf} \quad (42,000 \text{ lbf})$$

The answer is (C).

Why Other Options Are Wrong

(A) This incorrect option did not include the load from the concrete deck in the total dead load.

(B) This incorrect option included only the weight from one truss.

(D) This incorrect option calculated the total dead load, not the dead load on each abutment.

34. To determine which option is most cost effective, calculate and compare the labor hour for each option and the number of days required to complete construction.

Option A

If the roofing company uses its existing project team, the labor hour is $45, and the productivity rate is 32 ft^2/hr. With an 8 hr day, the number of working days required to complete the 10,000 ft^2 roof is

$$\frac{10,000 \text{ ft}^2}{\left(32 \frac{\text{ft}^2}{\text{hr}}\right)\left(8 \frac{\text{hr}}{\text{day}}\right)} = 39.06 \text{ days} \quad (40 \text{ days})$$

The total cost of using the existing six project team members is

$$(6)\left(45 \frac{\$}{\text{hr}}\right)\left(8 \frac{\text{hr}}{\text{day}}\right)(40 \text{ days}) = \$86,400$$

Option B

Adjust the existing team's labor hour to include the cost of two junior contractors. The labor hour for the eight person team is

$$\frac{(6)\left(45 \frac{\$}{\text{hr}}\right) + (2)\left(20 \frac{\$}{\text{hr}}\right)}{8} = \$38.75/\text{hr}$$

The number of days required to finish the project using option B is

$$\frac{10,000 \text{ ft}^2}{\left(40 \frac{\text{ft}^2}{\text{hr}}\right)\left(8 \frac{\text{hr}}{\text{day}}\right)} = 31.25 \text{ days} \quad (32 \text{ days})$$

The total cost of using option B is

$$(8)\left(38.75 \frac{\$}{\text{hr}}\right)\left(8 \frac{\text{hr}}{\text{day}}\right)(32 \text{ days}) = \$79,360$$

Option C

Adjust the existing team's labor hour to include the cost of the two senior contractors. The labor hour for the eight person team is

$$\frac{(6)\left(45 \frac{\$}{\text{hr}}\right) + (2)\left(50 \frac{\$}{\text{hr}}\right)}{8} = \$46.25/\text{hr}$$

The number of days required to finish the roof using option C is

$$\frac{10,000 \text{ ft}^2}{\left(50 \frac{\text{ft}^2}{\text{hr}}\right)\left(8 \frac{\text{hr}}{\text{day}}\right)} = 25 \text{ days}$$

The total cost of using option C is

$$(8)\left(46.25 \frac{\$}{\text{hr}}\right)\left(8 \frac{\text{hr}}{\text{day}}\right)(25 \text{ days}) = \$74,000$$

Option D

Option D contracts out the job to a project team of 10 members. The labor hour is $50/hr, and completing the project will take 20 working days. The total cost of using option D is

$$(10)\left(50 \frac{\$}{\text{hr}}\right)\left(8 \frac{\text{hr}}{\text{day}}\right)(20 \text{ days}) = \$80,000$$

Comparing the options, option C is the most cost-effective option that allows the roof to be completed ahead of schedule.

The answer is (C).

Why Other Options Are Wrong

(A) This incorrect option is less cost effective than option C.

(B) This incorrect option is less cost effective than option C.

(D) This incorrect option is less cost effective than option C.

35. Use the formula for the present worth of a sinking fund. Substitute the maximum monthly payment for the monthly payment value, A, and convert the annual interest rate to a monthly interest rate, i. n is the number of compounding payments per terms of loan.

$$P = A\left(\frac{(1+i)^n - 1}{i(1+i)^n}\right)$$

$$= (\$3000)\left(\frac{\left(1 + \frac{0.08}{12}\right)^{60} - 1}{\left(\frac{0.08}{12}\right)\left(1 + \frac{0.08}{12}\right)^{60}}\right)$$

$$= \$147{,}955 \quad (\$150{,}000)$$

The answer is (B).

Why Other Options Are Wrong

(A) This incorrect option did not convert the annual interest rate to a monthly interest rate.

(C) This incorrect option multiplied the monthly payment by the total number of payments instead of calculating the present value of total payments.

(D) This incorrect option calculated the future value instead of the present value.

36. Estimate the quantity of each material. The volume, V, of concrete required to construct 10 concrete pads is equal to 10 times the volume of one pad.

$$V_{\text{pad}} = L_1 L_2 t = (15 \text{ ft})(20 \text{ ft})\left(\frac{8 \text{ in}}{12 \frac{\text{in}}{\text{ft}}}\right)$$

$$= 200 \text{ ft}^3$$

$$V_{\text{concrete}} = 10 V_{\text{pad}} = (10)\left(\frac{200 \text{ ft}^3}{27 \frac{\text{ft}^3}{\text{yd}^3}}\right)$$

$$= 74.1 \text{ yd}^3$$

To find the total weight of reinforcing steel required, the quantities of no. 4 and no. 5 bars must be calculated. According to the problem's illustration, the reinforcing steel is placed every 12 in for both transverse and longitudinal reinforcement.

For longitudinal reinforcement, the number, n, of no. 4 bars required is

$$n_{\text{no. 4}} = \frac{L_1}{s} + 1 \text{ bar} = \frac{(15 \text{ ft})\left(12 \frac{\text{in}}{\text{ft}}\right)}{12 \frac{\text{in}}{\text{bar}}} + 1 \text{ bar}$$

$$= 16 \text{ bars}$$

For transverse reinforcement, the number of no. 5 bars required is

$$n_{\text{no. 5}} = \frac{L_2}{s} + 1 \text{ bar} = \frac{(20 \text{ ft})\left(12 \frac{\text{in}}{\text{ft}}\right)}{12 \frac{\text{in}}{\text{bar}}} + 1 \text{ bar}$$

$$= 21 \text{ bars}$$

The unit weight of steel, γ, is 490 lbf/ft^3. The weight per linear foot, w, for no. 4 rebar is

$$w_{\text{no. 4}} = \gamma_{\text{steel}} A = \gamma_{\text{steel}}\left(\frac{\pi d^2}{4}\right)$$

$$= \left(490 \frac{\text{lbf}}{\text{ft}^3}\right)\left(\frac{\pi\left(\frac{0.5 \text{ in}}{12 \frac{\text{in}}{\text{ft}}}\right)^2}{4}\right)$$

$$= 0.668 \text{ lbf/ft}$$

The weight per linear foot for no. 5 rebar is

$$w_{\text{no. 5}} = \gamma_{\text{steel}} A = \gamma_{\text{steel}}\left(\frac{\pi d^2}{4}\right)$$

$$= \left(490 \frac{\text{lbf}}{\text{ft}^3}\right)\left(\frac{\pi\left(\frac{0.625 \text{ in}}{12 \frac{\text{in}}{\text{ft}}}\right)^2}{4}\right)$$

$$= 1.04 \text{ lbf/ft}$$

The total weight of the rebar required for 10 concrete pads is

$$W_{\text{rebar}} = 10(w_{\text{no. 4}} L_2 n_{\text{no. 4}} + w_{\text{no. 5}} L_1 n_{\text{no. 5}})$$

$$= (10)\left(\begin{array}{l}\left(0.668 \frac{\text{lbf}}{\text{ft}}\right)\left(20 \frac{\text{ft}}{\text{bar}}\right)(16 \text{ bars}) \\ + \left(1.04 \frac{\text{lbf}}{\text{ft}}\right)\left(15 \frac{\text{ft}}{\text{bar}}\right)(21 \text{ bars})\end{array}\right)$$

$$= 5414 \text{ lbf}$$

Calculate the area, A, of the formwork that will be located along the perimeter of the 10 concrete pads.

$$A_{\text{formwork}} = n_{\text{pads}} h_{\text{formwork}}(2L_1 + 2L_2)$$

$$= (10)\left(\frac{8 \text{ in}}{12 \frac{\text{in}}{\text{ft}}}\right)((2)(15 \text{ ft}) + (2)(20 \text{ ft}))$$

$$= 467 \text{ ft}^2$$

Given the quantities of material to be used, the cost of materials is

$$C_{materials} = C_{concrete} V_{concrete} + C_{rebar} W_{rebar}$$
$$+ C_{formwork} A_{formwork}$$
$$= \left(200 \; \frac{\$}{yd^3}\right)(74.1 \; yd^3) + \left(3 \; \frac{\$}{lbf}\right)(5414 \; lbf)$$
$$+ \left(6 \; \frac{\$}{ft^2}\right)(467 \; ft^2)$$
$$= \$33,864$$

Calculate the labor cost, C, for each part by multiplying the labor cost per hour by the number of hours of labor needed. For the formwork,

$$C_{carpenters} = n_{carpenters} r_{carpenters} t_{formwork}$$
$$= n_{carpenters} r_{carpenters} \left(\frac{A_{formwork}}{r_{formwork}}\right)$$
$$= (3)\left(40 \; \frac{\$}{hr}\right)\left(\frac{467 \; ft^2}{7 \; \frac{ft^2}{hr}}\right)$$
$$= \$8006$$

For the reinforcement,

$$C_{ironworkers} = n_{ironworkers} r_{ironworkers} t_{rebar}$$
$$= n_{ironworkers} r_{ironworkers} \left(\frac{W_{rebar}}{r_{rebar}}\right)$$
$$= (3)\left(60 \; \frac{\$}{hr}\right)\left(\frac{5414 \; lbf}{90 \; \frac{lbf}{hr}}\right)$$
$$= \$10,828$$

For the concrete,

$$C_{concrete \; workers} = n_{concrete \; workers} r_{concrete \; workers} t_{concrete}$$
$$= n_{concrete \; workers} r_{concrete \; workers} \left(\frac{V_{concrete}}{r_{concrete}}\right)$$
$$= (2)\left(55 \; \frac{\$}{hr}\right)\left(\frac{74.1 \; yd^3}{1 \; \frac{yd^3}{hr}}\right)$$
$$= \$8151$$

The total labor cost is

$$C_{labor} = C_{carpenters} + C_{ironworkers} + C_{concrete \; workers}$$
$$= \$8006 + \$10,828 + \$8151$$
$$= \$26,985$$

The total material and labor cost is then

$$C_{total} = C_{materials} + C_{labor} = \$33,864 + \$26,985$$
$$= \$60,849 \quad (\$61,000)$$

The answer is (C).

Why Other Options Are Wrong

(A) This incorrect option calculated the material and labor cost of only one concrete pad.

(B) This incorrect option included only one worker in each trade when calculating labor costs.

(D) This incorrect option did not convert the volume of concrete from cubic feet to cubic yards before multiplying the volume by the unit cost.

37. Calculate the weight of the steel frame components, W, using the known load per foot, P, and the length, L, of the steel shapes.

The weight for the 250 steel joists is

$$W_{joist} = 250PL = (250)\left(36 \; \frac{lbf}{ft}\right)(20 \; ft)$$
$$= 180,000 \; lbf$$

The weight of the 200 steel beams is

$$W_{beam} = 200PL = (200)\left(55 \; \frac{lbf}{ft}\right)(25 \; ft)$$
$$= 275,000 \; lbf$$

The weight of the 100 columns is

$$W_{column} = 100PL = (100)\left(38 \; \frac{lbf}{ft}\right)(15 \; ft)$$
$$= 57,000 \; lbf$$

Therefore, the total weight of the steel frame components is

$$W_{total} = W_{joist} + W_{beam} + W_{column}$$
$$= 180,000 \; lbf + 275,000 \; lbf + 57,000 \; lbf$$
$$= 512,000 \; lbf$$

The total cost of steel material and fabrication, including the cost of connections, welding, and splices, is

$$C_{material} = W_{total}(C_{steel} + C_{fabrication})$$
$$+ (0.15)\left(W_{total}(C_{steel} + C_{fabrication})\right)$$
$$= (512,000 \; lbf)\left(2 \; \frac{\$}{lbf} + 0.5 \; \frac{\$}{lbf}\right)$$
$$+ (0.15)\left((512,000 \; lbf)\left(2 \; \frac{\$}{lbf} + 0.5 \; \frac{\$}{lbf}\right)\right)$$
$$= \$1,472,000$$

Find the time it will take to finish the project based on the productivity rate of the crew.

$$\frac{\begin{array}{l} 250 \text{ pieces steel} \\ + 200 \text{ pieces steel} \\ + 100 \text{ pieces steel} \end{array}}{35 \dfrac{\text{pieces steel}}{\text{day}}} = 15.7 \text{ days} \quad (16 \text{ days})$$

The equipment cost for the crane needed to complete the work is

$$C_{\text{equipment}} = \left(2000 \ \frac{\$}{\text{day}}\right)(16 \text{ days}) = \$32{,}000$$

The labor cost for the crew to work 16 eight-hour days at the provided labor costs is

$$C_{\text{labor}} = (16 \text{ days})\left(8 \ \frac{\text{hr}}{\text{day}}\right)\left(\begin{array}{l} C_{\text{engineer}} + C_{\text{foreman}} \\ + 4C_{\text{ironworker}} + C_{\text{operator}} \end{array}\right)$$

$$= (16 \text{ days})\left(8 \ \frac{\text{hr}}{\text{day}}\right)$$

$$\times \left(100 \ \frac{\$}{\text{hr}} + 70 \ \frac{\$}{\text{hr}} + (4)\left(65 \ \frac{\$}{\text{hr}}\right) + 50 \ \frac{\$}{\text{hr}}\right)$$

$$= \$61{,}440$$

The total cost of the construction project is

$$\begin{aligned} C_{\text{total}} &= C_{\text{material}} + C_{\text{equipment}} + C_{\text{labor}} \\ &= \$1{,}472{,}000 + \$32{,}000 + \$61{,}440 \\ &= \$1{,}565{,}440 \quad (\$1{,}570{,}000) \end{aligned}$$

The answer is (D).

Why Other Options Are Wrong

(A) This incorrect option did not include the cost of steel connections and welding in the cost of materials.

(B) This incorrect option multiplied the cost of labor by only 16 days, rather than 16 eight-hour days.

(C) This incorrect option accounted for the wages of only one ironworker, not four.

38. The total area of the brick partitions is

$$A_{\text{net}} = A_{\text{total,gross}}(100\% - \%_{\text{mortar}}) - A_{\text{total,windows}}$$

$$= \left(\begin{array}{l} (12 \text{ ft})(9 \text{ ft}) + (10 \text{ ft})(9 \text{ ft}) + (12 \text{ ft})(9 \text{ ft}) \\ + (10 \text{ ft})(9 \text{ ft}) \end{array}\right)$$

$$\times \left(\frac{100\% - 5\%}{100\%}\right) - (4)(3 \text{ ft})(4 \text{ ft})$$

$$= 328.2 \text{ ft}^2$$

The labor hour for the brick mason and the two assistants is

$$\frac{65 \ \dfrac{\$}{\text{hr}} + (2)\left(40 \ \dfrac{\$}{\text{hr}}\right)}{3} = \$48.33/\text{hr}$$

Calculate the time required to construct the brick walls.

$$\frac{328.2 \text{ ft}^2}{2 \ \dfrac{\text{ft}^2}{\text{hr}}} = 164.1 \text{ hr}$$

Therefore, the construction labor cost is

$$(3)\left(48.33 \ \frac{\$}{\text{hr}}\right)(164.1 \text{ hr}) = \$23{,}792.86$$

The same construction team must also repoint the mortar. Calculate the labor hour for the team.

$$\frac{65 \ \dfrac{\$}{\text{hr}} + 40 \ \dfrac{\$}{\text{hr}}}{2} = \$52.50/\text{hr}$$

The time required to repoint the mortar is

$$\frac{328.2 \text{ ft}^2}{3 \ \dfrac{\text{ft}^2}{\text{hr}}} = 109.4 \text{ hr}$$

Therefore, the repointing labor cost is

$$(2)\left(52.50 \ \frac{\$}{\text{hr}}\right)(109.4 \text{ hr}) = \$11{,}487$$

Calculate the material cost of brick.

$$\left(10 \ \frac{\$}{\text{ft}^2}\right)(328.2 \text{ ft}^2) = \$3282$$

The total material and labor cost is

$$\$3282 + \$23{,}792.85 + \$11{,}487 = \$38{,}561.85 \quad (\$39{,}000)$$

The answer is (C).

Why Other Options Are Wrong

(A) This incorrect option did not multiply the cost of labor by the number of crew members for either the construction or the repointing processes.

(B) This incorrect option did not multiply the cost of labor by the number of crew members for either the construction or the repointing processes, nor did it deduct the mortar area from the total area of the brick.

(D) This incorrect option did not deduct the mortar area from the total area of the brick.

39. Calculate the total cost of excavation and hauling by first finding the volume of soil to be excavated.

$$V_{\text{soil}} = Lwd = \frac{(200 \text{ ft})(50 \text{ ft})(10 \text{ ft})}{27 \ \dfrac{\text{ft}^3}{\text{yd}^3}}$$

$$= 3704 \text{ yd}^3$$

The total cost of excavation is

$$(3704 \text{ yd}^3)\left(12 \ \frac{\$}{\text{yd}^3}\right) = \$44{,}444$$

Find the number of trips the truck must take to haul all the excavated soil.

$$\frac{3704 \text{ yd}^3}{18 \ \dfrac{\text{yd}^3}{\text{trip}}} = 205.8 \text{ trips} \quad (206 \text{ trips})$$

The total cost of hauling is

$$(206 \text{ trips})\left(150 \ \frac{\$}{\text{trip}}\right) = \$30{,}900$$

Calculate the total costs for each crew.

Crew 1

Calculate the total cost to install the sheet pile. Find the total surface area of sheet pile needed by multiplying the length of the perimeter of the foundation by the height of the sheet pile needed. The sheet pile will extend 2 ft below and 3 ft above the height of the foundation, so the surface area is

$$\begin{aligned} A_{\text{pile}} &= L_{\text{perimeter}}h \\ &= (200 \text{ ft} + 50 \text{ ft} + 200 \text{ ft} + 50 \text{ ft}) \\ &\quad \times (2 \text{ ft} + 10 \text{ ft} + 3 \text{ ft}) \\ &= 7500 \text{ ft}^2 \end{aligned}$$

The material cost of the sheet pile is

$$\left(13 \ \frac{\$}{\text{ft}^2}\right)(7500 \text{ ft}^2) = \$97{,}500$$

Given the productivity rate of crew 1, the time required to install the sheet piles is

$$\frac{7500 \text{ ft}^2}{50 \ \dfrac{\text{ft}^2}{\text{hr}}} = 150 \text{ hr}$$

The labor hour for crew 1 is

$$\frac{\left(60 \ \dfrac{\$}{\text{hr}}\right)(2) + \left(30 \ \dfrac{\$}{\text{hr}}\right)(2)}{4} = \$45/\text{hr}$$

The total cost of labor for crew 1 is

$$(150 \text{ hr})\left(45 \ \frac{\$}{\text{hr}}\right)(4) = \$27{,}000$$

The number of days the sheet pile driver must be operated to complete all sheet pile work is

$$\frac{150 \text{ hr}}{8 \ \dfrac{\text{hr}}{\text{day}}} = 18.75 \text{ days} \quad (19 \text{ days})$$

The total equipment cost to complete the sheet pile work is

$$\left(1200 \ \frac{\$}{\text{day}}\right)(19 \text{ days}) = \$22{,}800$$

The total cost of the sheet pile work is

$$\$97{,}500 + \$27{,}000 + \$22{,}800 = \$147{,}300$$

Crew 2

Calculate the total cost of installing the wales and struts. 1000 ft of W24 × 76 struts weigh 76 lbf per foot of steel beam. So, the total weight of steel required is

$$(1000 \text{ ft})\left(76 \ \frac{\text{lbf}}{\text{ft}}\right) = 76{,}000 \text{ lbf}$$

The material cost of the steel beams is

$$\left(2.5 \ \frac{\$}{\text{lbf}}\right)(76{,}000 \text{ lbf}) = \$190{,}000$$

Given the productivity rate of crew 2, the time required to install the wales and struts is

$$\frac{1000 \text{ ft}}{20 \ \dfrac{\text{ft}}{\text{hr}}} = 50 \text{ hr}$$

The labor hour for crew 2 is

$$\frac{50 \ \frac{\$}{hr} + (2)\left(30 \ \frac{\$}{hr}\right)}{3} = \$36.67/hr$$

The labor cost for the placement of the wales and struts is

$$\left(36.67 \ \frac{\$}{hr}\right)(50 \ hr)(3) = \$5500.50$$

The total cost of placement of the wales and struts is

$$\$190,000 + \$5500.50 = \$195,500.50$$

Therefore, the total cost of the project is

$$\$44,444 + \$30,900 + \$147,300 + \$195,500$$
$$= \$418,144 \quad (\$420,000)$$

The answer is (D).

Why Other Options Are Wrong

(A) This incorrect option assumed the weight of steel required for the wales and struts was 1000 lbf instead of multiplying the steel's length by the steel's unit weight.

(B) This incorrect option calculated the sheet pile surface area, but neglected to include the extra depth beneath the excavation line and the height above the ground.

(C) This incorrect option estimated the labor cost by multiplying the labor hour by the time required to complete each procedure without multiplying by the number of workers in each crew.

40. The sum-of-the-years' digits for a five-year depreciation period is

$$5 + 4 + 3 + 2 + 1 = 15$$

The total depreciation is $\$20,000 - \$5000 = \$15,000$. The sheepsfoot roller is worth $\$20,000$ at the beginning of the first year. In the first year, the sheepsfoot roller will depreciate 5/15, or 33.33%, of the total depreciation. The depreciation, D_1, will be

$$D_1 = (\$15,000)(0.3333) = \$4995$$

At the beginning of the second year, the value of the sheepsfoot roller will be

$$\$20,000 - \$4995 = \$15,005$$

In the second year, the sheepsfoot roller will depreciate 4/15, or 26.67%, of the total depreciation. The depreciation, D_2, will be

$$D_2 = (\$15,000)(0.2667)$$
$$= \$4001$$

At the beginning of the third year, the value of the sheepsfoot roller will be

$$\$15,005 - \$4001 = \$11,004$$

In the third year, the sheepsfoot roller will depreciate 3/15, or 20%, of the total depreciation. The depreciation, D_3, will be

$$D_3 = (\$15,000)(0.2)$$
$$= \$3000$$

At the end of the third year, the value of the sheepsfoot roller will be

$$\$11,004 - \$3000 = \$8004 \quad (\$8000)$$

The answer is (B).

Why Other Options Are Wrong

(A) This incorrect option used the 200% declining balance depreciation method instead of the sum-of-the-years' digits method.

(B) This incorrect option used the straight line depreciation method instead of the sum-of-the-years' digits method.

(D) This incorrect option used the average depreciation rate method instead of the sum-of-the-years' digits method.

41. Calculate the amount deposited into the engineer's 401(k) savings account each month.

$$\text{monthly contribution} = \left(\frac{\text{annual income}}{12}\right)$$
$$\times \left(\begin{array}{c}\text{contribution} \\ \text{percentage}\end{array}\right)$$
$$= \left(\frac{\$70,000}{12}\right)(0.05)$$
$$= \$291.67$$

Calculate the monthly interest rate, given an 8% annual interest.

$$i = \frac{0.08}{12} = 0.0067$$

Rearrange the formula for the future worth of an annual amount to solve for n. Use the monthly contribution as the value of the monthly amount, A, and the monthly interest rate as the value of the interest rate, i, so that n will equal the number of monthly payments.

$$F = A\left(\frac{F}{A}, i, n\right) = A\left(\frac{(1+i)^n - 1}{i}\right)$$

$$\$1{,}000{,}000 = (\$291.67)\left(\frac{(1+0.0067)^n - 1}{0.0067}\right)$$

$$(1 + 0.0067)^n = 22.97$$

$$n \log 1.0067 = \log 23.97$$

$$n = 475.7$$

475.7 monthly payments must be made for the engineer to have \$1,000,000 in the 401(k). Convert this into years of payments required.

$$\frac{475.7 \text{ mo}}{12 \frac{\text{mo}}{\text{yr}}} = 39.6 \text{ yr} \quad (40 \text{ yr})$$

The engineer must work 40 years.

The answer is (C).

Why Other Options Are Wrong

(A) This incorrect option divided the annual amount into monthly increments but did not break the interest rate down into monthly interest.

(B) This incorrect option did not divide the annual amount into monthly increments.

(D) This incorrect option is the age at which the engineer will retire, not how many more years the engineer must work.

42. If the *actual cost of work performed* (ACWP) is greater than the *budgeted cost of work performed* (BCWP), the project is incurring greater costs than are allowed for in the budget. This means that the project is losing money, so option A is false.

If the *budgeted cost of worked scheduled* (BCWS) is less than the BCWP, the project is ahead of schedule, so option C is false.

The *cost performed index* (CPI) is found by dividing the BCWP by the ACWP. At the completion of the project, the BCWP was smaller than the ACWP, so dividing them will not produce a value greater than one. Option D is false.

From week three to week five, the BCWP value was greater than either the ACWP or the BCWS values, meaning the project was ahead of schedule and profitable during those weeks. Therefore, option B is the only correct statement regarding the project's performance.

The answer is (B).

Why Other Options Are Wrong

(A) This incorrect option assumed the BCWP at the end of the project was greater than the ACWP.

(C) This incorrect option assumed the BCWS was greater than the BCWP for the entire duration of the project. (The reverse was true during week 1 and week 2.)

(D) This incorrect option assumed the BCWP was greater than ACWP at the end of the project.

43. The tension, T, of sling 1 and sling 2 can be broken down into vertical and horizontal components. For sling 1,

$$T_{1,\text{vert}} = T_1 \sin 45°$$

$$T_{1,\text{horiz}} = T_1 \cos 45°$$

For sling 2,

$$T_{2,\text{vert}} = T_2 \sin 60°$$

$$T_{2,\text{horiz}} = T_2 \cos 60°$$

Since the system must be in static equilibrium, the horizontal forces from $T_{1,\text{horiz}}$ and $T_{2,\text{horiz}}$ must cancel each other. Therefore,

$$T_{1,\text{horiz}} = T_{2,\text{horiz}}$$

$$T_1 \cos 45° = T_2 \cos 60°$$

$$T_2 = \left(\frac{\cos 45°}{\cos 60°}\right) T_1$$

$$= 1.414 \, T_1$$

Because of the static equilibrium, the moment around any point in the system is zero. Take the moment around point C to find T_1. Assume counterclockwise moments are positive.

$$M_\text{C} = W_\text{A} d_\text{A} + W_\text{B} d_\text{B} - (T_1 \sin 45°) d_\text{A} = 0$$

$$T_1 = \frac{W_\text{A} d_\text{A} + W_\text{B} d_\text{B}}{(\sin 45°) d_\text{A}}$$

$$= \frac{(3000 \text{ lbf})(8 \text{ ft}) + (1500 \text{ lbf})(4 \text{ ft})}{(0.7071)(8 \text{ ft})}$$

$$= 5303 \text{ lbf}$$

The value of T_2 is

$$T_2 = 1.414 \, T_1 = (1.414)(5303 \text{ lbf})$$

$$= 7498 \text{ lbf}$$

Find the weight of object C, W_C, by taking the moment around point A. Assume counterclockwise moments are positive.

$$M_A = -W_B d_B - W_C d_C + (T_2 \sin 60°) d_C = 0$$

$$W_C = \frac{(T_2 \sin 60°) d_C - W_B d_B}{d_C}$$

$$= \frac{(7498 \text{ lbf})(0.8660)(8 \text{ ft}) - (1500 \text{ lbf})(4 \text{ ft})}{8 \text{ ft}}$$

$$= 5743 \text{ lbf} \quad (5700 \text{ lbf})$$

The answer is (C).

Why Other Options Are Wrong

(A) This incorrect option analyzed the system as though the slings were symmetrical.

(B) This incorrect option found the tension force in sling 1, not the weight of object C.

(D) This incorrect option found the tension force in sling 2, not the weight of object C.

44. If the horizontal center of gravity of each material (d_B and d_C, measured from point A) is multiplied by that material's weight (F_B and F_C), these two products added together will equal the horizontal center of gravity of the entire object (x_{cg}) multiplied by the weight of the entire object.

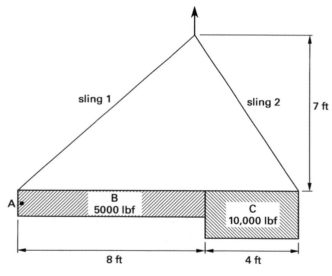

That is,

$$F_B d_B + F_C d_C = (F_B + F_C) x_{cg}$$

$$x_{cg} = \frac{F_B d_B + F_C d_C}{F_B + F_C}$$

$$= \frac{(5000 \text{ lbf})(4 \text{ ft}) + (10,000 \text{ lbf})(10 \text{ ft})}{5000 \text{ lbf} + 10,000 \text{ lbf}}$$

$$= 8 \text{ ft}$$

The object's horizontal center of gravity is 8 ft from point A, which is where the materials are connected. The master link should be aligned vertically. The length

of sling 1 can therefore be calculated using the Pythagorean theorem, with the center of gravity and the height of the system as the other two sides of the triangle.

$$L_1 = \sqrt{x_{cg}^2 + h^2} = \sqrt{(8 \text{ ft})^2 + (7 \text{ ft})^2}$$

$$= 10.63 \text{ ft} \quad (11 \text{ ft})$$

The answer is (C).

Why Other Options Are Wrong

(A) This incorrect option calculated the length of sling 2.

(B) This incorrect option assumed the center of gravity is in the middle of the system.

(D) This incorrect option calculated the total length of sling 1 and sling 2.

45. The total weight the crane must lift is

$$15,000 \text{ lbf} + 2000 \text{ lbf} = 17,000 \text{ lbf}$$

From the table provided, the crane can safely lift 15,940 lbf at a 170 ft lift radius, and 18,440 lbf at a 150 ft lift radius. To calculate the exact maximum lift radius at which the crane can safely lift 17,000 lbf, interpolation is needed.

$$\frac{18,440 \text{ lbf} - 15,940 \text{ lbf}}{150 \text{ ft} - 170 \text{ ft}} = \frac{17,000 \text{ lbf} - 15,940 \text{ lbf}}{r - 170 \text{ ft}}$$

Interpolating, the maximum lift radius at which the crane can safely lift 17,000 lbf is 161 ft (160 ft).

The answer is (B).

Why Other Options Are Wrong

(A) This incorrect option did not interpolate the maximum lift radius.

(C) This incorrect option measured the maximum lift radius for a 15,000 lbf load.

(D) This incorrect option measured the maximum lift radius for the load from lifting accessories only.

46. The crane system needs to be in static equilibrium, so the total moment at any point is zero. Therefore, the turning (clockwise) and counter (counterclockwise) moments taken at the front end of the crane body should cancel. Find the turning moment.

$$M_{\text{turning}} = W_{\text{truss}} d_{\text{truss}} + W_{\text{boom}} d_{\text{boom}}$$

$$= (5 \text{ ton})(30 \text{ ft}) + (3 \text{ ton})(15 \text{ ft})$$

$$= 195 \text{ ft-tons}$$

Find the unweighted counter moment from the front end of the crane body to the midpoint of the crane body.

$$M_{\text{counter,unweighted}} = W_{\text{body}} d_{\text{body}} = (28 \text{ ton})(10 \text{ ft})$$
$$= 280 \text{ ft-tons}$$

If the quotient of the counter moment and the turning moment is less than 1.5, a counterweight must be added.

$$\frac{M_{\text{counter,unweighted}}}{M_{\text{turning}}} = \frac{280 \text{ ft-tons}}{195 \text{ ft-tons}}$$
$$= 1.44 \quad [< 1.5, \text{ counterweight needed}]$$

The counter moment must be at least 1.5 times the turning moment. Find the tonnage of the counterweight.

$$M_{\text{counter,weighted}} = M_{\text{counter,unweighted}} + W_{\text{counter}} d_{\text{counter}}$$
$$= 1.5 M_{\text{turning}}$$
$$W_{\text{counter}} = \frac{1.5 M_{\text{turning}} - M_{\text{counter,unweighted}}}{d_{\text{counter}}}$$
$$= \frac{(1.5)(195 \text{ ft-tons}) - 280 \text{ ft-tons}}{18 \text{ ft}}$$
$$= 0.694 \text{ tons} \quad (0.7 \text{ tons})$$

The crane requires a counterweight of 0.7 tons.

The answer is (A).

Why Other Options Are Wrong

(B) This incorrect option incorrectly assumed the center of gravity of the boom at the lifting hook.

(C) This incorrect option did not take into account the effect of turning moments and assumed the crane would work solely based on weight.

(D) This incorrect option has a counter moment smaller than 1.5 times the turning moment even if boom weight is not included.

47. Since the well's radius and hydraulic conductivity are given, the depth of the water in the well, y_2, may be found by rearranging the Dupuit equation.

$$Q = \frac{\pi K(y_2^2 - y_1^2)}{\ln \frac{r_2}{r_1}}$$

$$y_1 = \sqrt{y_2^2 - \frac{Q \ln \frac{r_2}{r_1}}{\pi K}}$$

$$= \sqrt{(200 \text{ ft})^2 - \frac{\left(\dfrac{\left(80 \, \frac{\text{gal}}{\text{min}}\right)\left(1440 \, \frac{\text{min}}{\text{day}}\right)}{7.48 \, \frac{\text{gal}}{\text{ft}^3}}\right) \ln \frac{1500 \text{ ft}}{1 \text{ ft}}}{\pi \left(1.0 \, \frac{\text{ft}}{\text{day}}\right)}}$$

$$= 64.4 \text{ ft} \quad (65 \text{ ft})$$

The answer is (C).

Why Other Options Are Wrong

(A) This incorrect option assumed that the influence of the pump would be linear and that the level of water at the well is zero.

(B) This incorrect option used the diameter of the unconfined acquifer instead of the radius, and incorrectly calculated the level of water.

(D) This incorrect option used the diameter of the well instead of the radius and incorrectly calculated the level of water.

48. The horsepower of the pump is the product of the quantity of water and the total head required by the system. Use the extended Bernoulli equation.

$$(E_p + E_v + E_z)_1 + E_A = (E_p + E_v + E_z)_2 + E_E$$
$$+ E_f + E_m$$

As specific energy can be translated into head, the following equation can be obtained.

$$(h_p + h_v + h_z)_1 + h_A = (h_p + h_v + h_z)_2 + h_E + h_f + h_m$$

Pressure head, h_p, and velocity head, h_v, do not change between the reservoirs, so these variables can be canceled from both sides of the equation. No head is lost due to turbine extraction, h_E, and minor losses, h_m, can be neglected according to the problem, so this equation becomes

$$h_{z,1} + h_A = h_{z,2} + h_f$$

Find the friction losses. A conversion factor of $2.31 \text{ ft/(lbf/in}^2)$ is used to convert pressure to head.

$$h_f = 5 \text{ ft} + 7 \text{ ft} + \left(15 \, \frac{\text{lbf}}{\text{in}^2}\right)\left(2.31 \, \frac{\text{ft}}{\frac{\text{lbf}}{\text{in}^2}}\right)$$
$$+ \left(10 \, \frac{\text{lbf}}{\text{in}^2}\right)\left(2.31 \, \frac{\text{ft}}{\frac{\text{lbf}}{\text{in}^2}}\right)$$
$$= 69.75 \text{ ft}$$

The head that must be added by the pump is then

$$h_A = (h_{z,2} - h_{z,1}) + h_f = 500 \text{ ft} + 69.75 \text{ ft} = 569.75 \text{ ft}$$

The minimum water (hydraulic) horsepower needed is

$$\text{WHP} = \frac{h_A Q(\text{SG})}{3956} = \frac{(569.75 \text{ ft})\left(500 \, \frac{\text{gal}}{\text{min}}\right)(1.0)}{3956 \, \frac{\text{ft-gal}}{\text{hp-min}}}$$
$$= 72.01 \text{ hp} \quad (70 \text{ hp})$$

The answer is (C).

Alternative Solution

An alternative method to solve this problem is to use the formula

$$\text{WHP} = \frac{h_A \dot{m}}{550} \times \frac{g}{g_c}$$

The volumetric flow rate is

$$\dot{V} = Q = \frac{500 \ \dfrac{\text{gal}}{\text{min}}}{\left(7.48 \ \dfrac{\text{gal}}{\text{ft}^3}\right)\left(60 \ \dfrac{\text{sec}}{\text{min}}\right)}$$

$$= 1.11 \ \text{ft}^3/\text{sec}$$

Converting to mass flow rate gives

$$\dot{m} = \rho \dot{V} = \left(62.4 \ \frac{\text{lbm}}{\text{ft}^3}\right)\left(1.11 \ \frac{\text{ft}^3}{\text{sec}}\right)$$

$$= 69.3 \ \text{lbm/sec}$$

The water (hydraulic) horsepower required is

$$\text{WHP} = \frac{h_A \dot{m}}{550} \times \frac{g}{g_c}$$

$$= \frac{(569.75 \ \text{ft})\left(69.3 \ \dfrac{\text{lbm}}{\text{sec}}\right)}{550 \ \dfrac{\text{ft-lbf}}{\text{hp-sec}}} \times \frac{32.2 \ \dfrac{\text{ft}}{\text{sec}^2}}{32.2 \ \dfrac{\text{lbm-ft}}{\text{lbf-sec}^2}}$$

$$= 71.78 \ \text{hp} \quad (70 \ \text{hp})$$

The answer is (C).

Why Other Options Are Wrong

(A) This incorrect option used the volumetric flow rate (in ft^3/sec) instead of the mass flow rate (in lbm/sec) when solved using the alternative method.

(B) This incorrect option neglected all friction heads.

(D) This incorrect option used the volumetric flow rate (in gal/min) instead of the mass flow rate (in lbm/sec) when solved using the alternative method.

49. Taking the cycle time and swing-depth factor from Table 49.1 and Table 49.2 given in the problem, the production of the excavator is

$$P = (\text{cycle time})(\text{swing-depth factor}) V_{\text{bucket}}$$

$$\times (\text{bucket factor})(\text{efficiency})$$

$$= \left(160 \ \frac{\text{cycles}}{\text{hr}}\right)(1.10)\left(2 \ \frac{\text{yd}^3}{\text{cycle}}\right)(0.9)\left(\frac{50 \ \text{min}}{60 \ \text{min}}\right)$$

$$= 264 \ \text{yd}^3/\text{hr} \quad (260 \ \text{yd}^3/\text{hr})$$

The answer is (B).

Why Other Options Are Wrong

(A) This incorrect option did not include the swing depth factor in the calculation.

(C) This incorrect option did not include the bucket factor in the calculation.

(D) This incorrect option did not include the work efficiency in the calculation.

50. Convert the production, P, of the static compact roller into cubic yards per hour. w is the width of the roller, v is the speed of the roller, B is the depth of the lift, and n is the number of passes.

$$P = \frac{wvB}{n} = \frac{(8 \ \text{ft})\left(\left(5 \ \dfrac{\text{mi}}{\text{hr}}\right)\left(5280 \ \dfrac{\text{ft}}{\text{mi}}\right)\right)\left(\dfrac{24 \ \text{in}}{12 \ \dfrac{\text{in}}{\text{ft}}}\right)}{(2)\left(27 \ \dfrac{\text{ft}^3}{\text{yd}^3}\right)}$$

$$= 7822 \ \text{yd}^3/\text{hr}$$

The total volume to be compacted is

$$V = Ldh = \frac{(200 \ \text{ft})(300 \ \text{ft})(2 \ \text{ft})(2)}{27 \ \dfrac{\text{ft}^3}{\text{yd}^3}}$$

$$= 8889 \ \text{yd}^3$$

Therefore, the time that it requires to compact two lifts of soil is

$$t_{\text{required}} = \frac{V}{P} = \frac{8889 \ \text{yd}^3}{7822 \ \dfrac{\text{yd}^3}{\text{hr}}}$$

$$= 1.14 \ \text{hr} \quad (1.2 \ \text{hr})$$

The answer is (D).

Why Other Options Are Wrong

(A) This incorrect option did not convert the depth of the lift from inches to feet when calculating the production of the roller per hour.

(B) This incorrect option did not divide the production volume by 2 to account for the second pass. Furthermore, it included only one lift.

(C) This incorrect option did not divide the production volume by 2 to account for the second pass.

51. From the table, the cycle time of a single pusher using the chain method is 1 min. The number of scrapers, n, one pusher can serve in a cycle, t, is

$$n = \frac{t_{\text{scraper}}}{t_{\text{pusher}}} = \frac{5 \ \text{min}}{1 \ \text{min}} = 5$$

Find the number of pushers required to serve eight scrapers.

$$n_{\text{required}} = \frac{8}{5} = 1.6 \quad (2)$$

The answer is (B).

Why Other Options Are Wrong

(A) This incorrect option rounded down the number of pushers required rather than rounding up.

(C) This incorrect option only solved for the number of scrapers one pusher can serve.

(D) This incorrect option assumed that one pusher services each scraper separately.

52. Plot the load growth data.

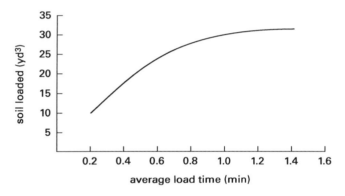

As load time increases, the curve eventually plateaus. Extend the x-axis to the left to find a point on the x-axis where the distance to the origin is the cycle time, 2 min. This point is marked as point A.

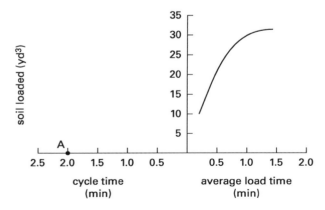

Draw a straight line from point A tangent to the load growth curve. The tangent point, point D, gives the optimal load time, point B, and load volume, point C.

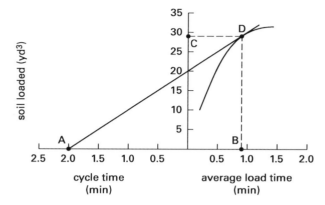

Plotting this tangent line is used to find the optimal production rate. The production rate is the soil loaded divided by the average load time, so the line with the highest slope will equal the optimal rate.

The positions of point A and the curve are fixed by the data. The distance AB represents the total cycle time of the scraper and distance BD represents the load volume per cycle. Therefore, the slope of line AD represents the rate of production. This will be at its maximum when point D is chosen so that line AD is tangent to the curve.

Therefore, the optimal loading time is 0.9 min, and the corresponding optimum load is 29 yd^3. The total cycle time for the scraper is

$$2 \text{ min} + 0.9 \text{ min} = 2.9 \text{ min} \quad (3.0 \text{ min})$$

The answer is (C).

Why Other Options Are Wrong

(A) This incorrect option neglected to include the 2 min cycle time.

(B) This incorrect option calculated the loading time for when the greatest volume of material was loaded and also neglected to include the 2 min cycle time.

(D) This incorrect option calculated the loading time for when the greatest volume of material was loaded, which is not the same as the optimal loading time.

53. *Sheet piles* are rolled steel sections driven into unexcavated ground before trenching begins. They are placed side by side and usually have interlocking edges. When locked together, they act as a permanent retaining wall.

Temporary control measures used during excavation include silt fences, slope drains, and brush barriers. *Silt fences* attach woven wire or other woven material to posts with filter cloth. They are used to protect water quality of nearby water bodies from construction runoff water. *Slope drains* (e.g., fiber mats, plastic sheets, and concrete gutters) carry water down slopes. They are used prior to the installation of permanent structures.

Brush barriers are brush, tree linings, shrubs, or other plants that are placed at the bottom of fill slopes to restrain sedimentation particles.

The answer is (A).

Why Other Options Are Wrong

(B) This incorrect option is a temporary erosion control measure used during excavation.

(C) This incorrect option is a temporary erosion control measure used during excavation.

(D) This incorrect option is a temporary erosion control measure used during excavation.

54. The total cost of the pipe is

$$C_{\text{pipe}} = (200 \text{ ft})\left(30 \ \frac{\$}{\text{ft}}\right)$$
$$= \$6000$$

Find the volume of gravel needed to construct the base.

$$V_{\text{gravel}} = w_{\text{trench}} L_{\text{trench}} D_{\text{base}} = (3 \text{ ft})(200 \text{ ft})(1 \text{ ft})$$
$$= 600 \text{ ft}^3$$

The total cost of the gravel is

$$C_{\text{gravel}} = \left(\frac{600 \text{ ft}^3}{27 \ \frac{\text{ft}^3}{\text{yd}^3}}\right)\left(36 \ \frac{\$}{\text{yd}^3}\right)$$
$$= \$800$$

Find the volume of backfill needed to fill the trench. The volume of backfill will be equal to the total volume of the trench minus the volume of the base and the volume of the pipe.

$$V_{\text{trench}} = wLD = (3 \text{ ft})(6 \text{ ft})(200 \text{ ft})$$
$$= 3600 \text{ ft}^3$$
$$V_{\text{pipe}} = \pi r^2 L = \pi (0.5 \text{ ft})^2 (200 \text{ ft})$$
$$= 157.08 \text{ ft}^3$$
$$V_{\text{backfill}} = V_{\text{trench}} - V_{\text{pipe}} - V_{\text{gravel}}$$
$$= 3600 \text{ ft}^3 - 157.08 \text{ ft}^3 - 600 \text{ ft}^3$$
$$= 2842.92 \text{ ft}^3 \quad (2843 \text{ ft}^3)$$

Because the borrow pit soil is unlikely to be purchased in fractions of a cubic yard, use the rounded value of 2843 ft³. The total cost of the backfill is

$$C_{\text{backfill}} = \left(\frac{2843 \text{ ft}^3}{27 \ \frac{\text{ft}^3}{\text{yd}^3}}\right)\left(25 \ \frac{\$}{\text{yd}^3}\right)$$
$$= \$2632.41$$

Calculate the quantity and cost of the timber.

The number of studs on one side of the trench is

$$N_{\text{studs}} = \frac{L_{\text{trench}}}{\text{spacing of studs}} = \frac{200 \text{ ft}}{6 \text{ ft}}$$
$$= 33.3 \text{ studs} \quad (34 \text{ studs})$$

The studs are placed on either side of the trench, so the total number of studs is 68.

Find the board foot measure for the studs, wales, and cross bracing. (A board foot (bd-ft) is equal to 144 in³.)

For the 68 studs,

$$N_{\text{studs}} L_{\text{stud}} A_{\text{stud}} = \frac{(68)(6 \text{ ft})\left(12 \ \frac{\text{in}}{\text{ft}}\right)(4 \text{ in})(8 \text{ in})}{144 \ \frac{\text{in}^3}{\text{bd-ft}}}$$
$$= 1088 \text{ bd-ft}$$

For the four wales at 200 ft each,

$$N_{\text{wales}} L_{\text{wale}} A_{\text{wale}} = \frac{(4)(200 \text{ ft})\left(12 \ \frac{\text{in}}{\text{ft}}\right)(2 \text{ in})(6 \text{ in})}{144 \ \frac{\text{in}^3}{\text{bd-ft}}}$$
$$= 800 \text{ bd-ft}$$

For the cross bracing, first find the number of bracings by dividing the total length of trench by the spacing of studs, then find the board foot measure for the cross bracing.

$$N_{\text{bracings}} = \frac{L_{\text{trench}}}{s_{\text{studs}}} = \frac{200 \text{ ft}}{6 \text{ ft}} = 33.3 \quad (34)$$

There are two wale locations, so use 68 cross braces.

$$N_{\text{bracings}} L_{\text{bracing}} A_{\text{bracing}} = \frac{\begin{array}{c}(68)\left((3 \text{ ft})\left(12 \ \frac{\text{in}}{\text{ft}}\right)\right)\\ \times \left((2 \text{ in})(4 \text{ in})\right)\end{array}}{144 \ \frac{\text{in}^3}{\text{bd-ft}}}$$
$$= 136 \text{ bd-ft}$$

Therefore, the total board foot measure of the timber is

$$1088 \text{ bd-ft} + 800 \text{ bd-ft} + 136 \text{ bd-ft} = 2024 \text{ bd-ft}$$

The cost of the timber is \$4.00 per board foot.

$$\left(4.00 \ \frac{\$}{\text{bd-ft}}\right)(2024 \ \text{bd-ft}) = \$8096$$

The total cost of the materials is then

$$\$6000 + \$800 + \$2632 + \$8096 = \$17{,}528 \quad (\$17{,}500)$$

The answer is (B).

Why Other Options Are Wrong

(A) This incorrect option incorrectly estimated the amount of timber by including only one side of studs, one wale per side, and one cross bracing every 6 ft.

(C) This incorrect option did not subtract the pipe volume when estimating the quantity of backfill.

(D) This incorrect option did not subtract the gravel volume when estimating the quantity of backfill.

55. All activities on the critical path will have a total float of zero.

To find the total float of each activity, the earliest start (ES), earliest finish (EF), latest start (LS), and latest finish (LF) must be determined.

Determine the ES and EF. Calculate the earliest possible time that an activity can start after the previous activity finishes. The EF time will be the ES time plus the duration, D, of the activity. If an activity has more than one precedent, the earliest start time must be after whichever precedent activity finishes latest. That is,

$$\text{ES} = \text{EF}_{\text{predecessor}}$$
$$\text{EF} = \text{ES} + D$$

Determine the LS and LF. Calculate the latest possible time that an activity can start and finish without delaying the next activity or the project. To find the LS and LF time, use the backward path method. Because the last activity cannot be finished any later, the LS and LF dates will be the same as the ES and EF dates. By tracing the path backward to the activity preceding the last activity, the LF can be found because it is the LS of the last activity. Then, by subtracting the duration of the precedent activity, the LS date of the precedent activity can be found. By working backward, all other LS and LF dates can be found. That is,

$$\text{LF} = \text{LS}_{\text{successor}}$$
$$\text{LS} = \text{LF} - D$$

If an activity precedes more than one activity, the LF must be before the LS of any following activity.

The total float (TF) is the number of days that the activity can be delayed without delaying the entire project. If an activity has a TF value of zero, the activity is on the critical path.

The ES, EF, LS, LF, and total float are shown for each activity.

activity	ES (day)	EF (day)	LS (day)	LF (day)	TF (day)
AE	1	5	23	27	22
AD	1	8	9	16	8
AB	1	11	1	11	0
BC	11	16	11	16	0
CD	16	16	16	16	0
DE	16	27	16	27	0
BF	11	15	25	29	14
CF	16	19	26	29	10
EG	27	37	27	37	0
FG	19	27	29	37	10
GH	37	46	37	46	0

Therefore, the critical path is A-B-C-D-E-G-H.

The answer is (D).

Why Other Options Are Wrong

(A) This incorrect option included float time.

(B) This incorrect option included float time.

(C) This incorrect option included float time.

56. To find the total float of each activity, the earliest start (ES), earliest finish (EF), latest start (LS), and latest finish (LF) must be determined.

Determine the ES and EF. Calculate the earliest possible time that an activity can start after the previous activity finishes. The EF time will be the ES time plus the duration, D, of the activity. If an activity has more than one precedent, the ES time must be after the activity finishing latest. That is,

$$\text{ES} = \text{EF}_{\text{predecessor}}$$
$$\text{EF} = \text{ES} + D$$

Determine the LS and LF. Calculate the latest possible time that an activity can start and finish without delaying the next activity or the project. To find the LS and LF time, the path of the project has to be worked backward from the next activity's LS time to obtain the LF of the desired activity (or activities). The LS will be the LF date minus the duration of the activity. That is,

$$\text{LF} = \text{LS}_{\text{successor}}$$
$$\text{LS} = \text{LF} - D$$

The total float (TF) is

$$\text{TF} = \text{LS} - \text{ES} = \text{LF} - \text{EF}$$

The ES, EF, LS, LF, and total float are shown for each activity.

activity	ES (day)	EF (day)	LS (day)	LF (day)	TF (day)
AB	1	5	1	5	0
BC	5	13	5	13	0
BD	5	8	15	18	10
CD	13	18	13	18	0
DE	18	22	18	22	0
CF	13	19	25	31	12
EF	22	31	22	31	0
EG	22	28	32	38	10
FG	31	38	31	38	0

Therefore, the total float of activity BD is 10 days.

The answer is (C).

Why Other Options Are Wrong

(A) This incorrect option assumed the total float was zero.

(B) This incorrect option calculated the duration of activity BD, not the total float.

(D) This incorrect option calculated the total float for activity CF.

57. Perform a forward pass to obtain the earliest start (ES) and the earliest finish (EF) of each activity. An activity may begin when all activities leading to it (predecessors) are completed. If an activity has multiple predecessors, the ES value will be the largest EF value of all its predecessors. Then the activity's EF can be calculated as the sum of the activity's ES and its duration, D. That is, $EF = ES + D$. The start time is zero for relative calculations.

For activity A,

$$ES_A = 0 \text{ days}$$
$$EF_A = ES_A + D_A$$
$$= 0 \text{ days} + 9 \text{ days}$$
$$= 9 \text{ days}$$

For activity B,

$$ES_B = 0 \text{ days}$$
$$EF_B = ES_B + D_B$$
$$= 0 \text{ days} + 6 \text{ days}$$
$$= 6 \text{ days}$$

For activity C, the ES value will be the larger EF value of its two predecessors, activity A and activity B.

$$ES_C = EF_A = 9 \text{ days}$$
$$EF_C = ES_C + D_C$$
$$= 9 \text{ days} + 11 \text{ days}$$
$$= 20 \text{ days}$$

For activity D,

$$ES_D = EF_B = 6 \text{ days}$$
$$EF_D = ES_D + D_D$$
$$= 6 \text{ days} + 13 \text{ days}$$
$$= 19 \text{ days}$$

For activity E,

$$ES_E = EF_A = 9 \text{ days}$$
$$EF_E = ES_E + D_E$$
$$= 9 \text{ days} + 6 \text{ days}$$
$$= 15 \text{ days}$$

For activity F, the ES value will be the larger EF value of its two predecessors, activity C and activity D.

$$ES_F = EF_C = 20 \text{ days}$$
$$EF_F = ES_F + D_F$$
$$= 20 \text{ days} + 8 \text{ days}$$
$$= 28 \text{ days}$$

For activity G, the ES value will be the larger EF value of its two predecessors, activity E and activity F.

$$ES_G = EF_F = 28 \text{ days}$$
$$EF_G = ES_G + D_G$$
$$= 28 \text{ days} + 7 \text{ days}$$
$$= 35 \text{ days}$$

Perform a backward pass to find the latest start (LS) and latest finish (LF) of each activity. If an activity has multiple successors, the activity's LF will be the lower LS value of all its successors. Then the activity's LS can be calculated by subtracting the duration of the activity from its LF. That is, $LS = LF - D$.

Because G is the final activity, its EF and LF values are equal.

$$LF_G = EF_G = 35 \text{ days}$$

$$LS_G = LF_G - D_G$$

$$= 35 \text{ days} - 7 \text{ days}$$

$$= 28 \text{ days}$$

For activity F,

$$LF_F = LS_G = 28 \text{ days}$$

$$LS_F = LF_F - D_F$$

$$= 28 \text{ days} - 8 \text{ days}$$

$$= 20 \text{ days}$$

For activity E,

$$LF_E = LS_G = 28 \text{ days}$$

$$LS_E = LF_E - D_E$$

$$= 28 \text{ days} - 6 \text{ days}$$

$$= 22 \text{ days}$$

For activity D,

$$LF_D = LS_F = 20 \text{ days}$$

$$LS_D = LF_D - D_D$$

$$= 20 \text{ days} - 13 \text{ days}$$

$$= 7 \text{ days}$$

For activity C,

$$LF_C = LS_F = 20 \text{ days}$$

$$LS_C = LF_C - D_C$$

$$= 20 \text{ days} - 11 \text{ days}$$

$$= 9 \text{ days}$$

For activity B, the LF value will be the lower LS value of all its successors, activity C and activity D.

$$LF_B = LS_D = 7 \text{ days}$$

$$LS_B = LF_B - D_B$$

$$= 7 \text{ days} - 6 \text{ days}$$

$$= 1 \text{ day}$$

For activity A, the LF value will be the lower LS value of all its successors, activity C and activity E,

$$LF_A = LS_C = 9 \text{ days}$$

$$LS_A = LF_A - D_A$$

$$= 9 \text{ days} - 9 \text{ days}$$

$$= 0 \text{ days}$$

Calculate the total float (TF) for each activity. The TF is the amount of available extra time to complete an activity without delaying the project's completion date and is calculated as $TF = LF - EF = LS - ES$. The TF is always zero for the first and last activities.

For activity A,

$$TF_A = LS_A - ES_A$$

$$= 0 \text{ days} - 0 \text{ days}$$

$$= 0 \text{ days}$$

For activity B,

$$TF_B = LS_B - ES_B$$

$$= 1 \text{ day} - 0 \text{ days}$$

$$= 1 \text{ day}$$

For activity C,

$$TF_C = LS_C - ES_C$$

$$= 9 \text{ days} - 9 \text{ days}$$

$$= 0 \text{ days}$$

For activity D,

$$TF_D = LS_D - ES_D$$

$$= 7 \text{ days} - 6 \text{ days}$$

$$= 1 \text{ day}$$

For activity E,

$$TF_E = LS_E - ES_E$$

$$= 22 \text{ days} - 9 \text{ days}$$

$$= 13 \text{ days}$$

For activity F,

$$TF_F = LS_F - ES_F$$

$$= 20 \text{ days} - 20 \text{ days}$$

$$= 0 \text{ days}$$

For activity G,

$$TF_G = LS_G - ES_G$$

$$= 28 \text{ days} - 28 \text{ days}$$

$$= 0 \text{ days}$$

Complete the precedence diagram using the calculated values.

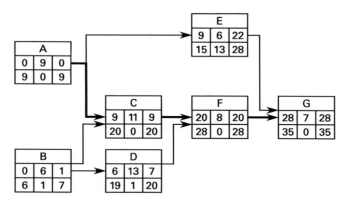

The TF for a project's critical path is always zero. Since A-C-F-G has a TF of zero, it is the critical path.

The answer is (B).

Why Other Options Are Wrong

(A) This incorrect option is not the critical path.

(C) This incorrect option is not the critical path.

(D) This incorrect option is not the critical path.

58. The free float (FF) of an activity is the number of days that an activity can be delayed without affecting the earliest start (ES) of a successor. It is the difference between the successor's ES and the earliest finish (EF) of the activity. Therefore, the FF of this activity is

$$FF_{activity} = ES_{successor} - EF_{activity}$$
$$= 34 \text{ days} - 28 \text{ days}$$
$$= 6 \text{ days}$$

However, when an activity has a lag time, L, the lag time must be subtracted from the free float value.

Therefore, the FF time of the activity is

$$FF - L = 6 \text{ days} - 2 \text{ days}$$
$$= 4 \text{ days}$$

The answer is (C).

Why Other Options Are Wrong

(A) This incorrect option calculated the free float of the predecessor activity with the inclusion of lag time.

(B) This incorrect option calculated the free float of the predecessor activity without including lag time.

(D) This incorrect option did not take into account the lag time.

59. In an activity-on-node diagram, each activity is represented by a node with an associated description and expected duration. For each node, the fraction's numerator represents the activity description, and the denominator represents the activity duration. Therefore, the duration of activity A is 2 days, and the duration of activity B is 6 days.

The answer is (A).

Why Other Options Are Wrong

(B) This incorrect option miscalculated the duration of activity B, which is 6 days.

(C) This incorrect option miscalculated the earliest start for activity A, which is 0 days.

(D) This incorrect option miscalculated the earliest finish for activity B, which is 8 days.

60. Draw the activity-on-node network from the parameters given in the problem statement.

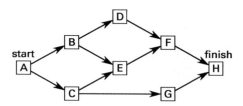

Perform a forward pass to calculate the earliest start (ES) and the earliest finish (EF) values for each activity.

For activity A, the ES value is 0 days because it is the first activity. The EF is then

$$EF_A = ES_A + D_A = 0 \text{ days} + 3 \text{ days}$$
$$= 3 \text{ days}$$

For activity B,

$$ES_B = EF_A = 3 \text{ days}$$
$$EF_B = ES_B + D_B = 3 \text{ days} + 5 \text{ days}$$
$$= 8 \text{ days}$$

For activity C,

$$ES_C = EF_A = 3 \text{ days}$$
$$EF_C = ES_C + D_C = 3 \text{ days} + 6 \text{ days}$$
$$= 9 \text{ days}$$

For activity D,

$$ES_D = EF_B = 8 \text{ days}$$
$$EF_D = ES_D + D_D = 8 \text{ days} + 3 \text{ days}$$
$$= 11 \text{ days}$$

For activity E, the ES value will be the larger EF value of its two predecessors, activity B and activity C.

$$ES_E = EF_C = 9 \text{ days}$$
$$EF_E = ES_E + D_E = 9 \text{ days} + 6 \text{ days}$$
$$= 15 \text{ days}$$

For activity F, the ES value will be the larger EF value of its two predecessors, activity D and activity E.

$$ES_F = EF_E = 15 \text{ days}$$
$$EF_F = ES_F + D_F = 15 \text{ days} + 7 \text{ days}$$
$$= 22 \text{ days}$$

For activity G,

$$ES_G = EF_C = 9 \text{ days}$$
$$EF_G = ES_G + D_G = 9 \text{ days} + 2 \text{ days}$$
$$= 11 \text{ days}$$

For activity H, the ES value will be the larger EF value of its two predecessors, activity F and activity G.

$$ES_H = EF_F = 22 \text{ days}$$
$$EF_H = ES_H + D_H = 22 \text{ days} + 3 \text{ days}$$
$$= 25 \text{ days}$$

Perform a backward pass to find the latest start (LS) and latest finish (LF) of each activity.

For activity H, its EF and LF values are equal because it is the final activity.

$$LF_H = EF_H = 25 \text{ days}$$
$$LS_H = LF_H - D_H = 25 \text{ days} - 3 \text{ days}$$
$$= 22 \text{ days}$$

For activity G,

$$LF_G = LS_H = 22 \text{ days}$$
$$LS_G = LF_G - D_G = 22 \text{ days} - 2 \text{ days}$$
$$= 20 \text{ days}$$

For activity F,

$$LF_F = LS_H = 22 \text{ days}$$
$$LS_F = LF_F - D_F = 22 \text{ days} - 7 \text{ days}$$
$$= 15 \text{ days}$$

For activity E,

$$LF_E = LS_F = 15 \text{ days}$$
$$LS_E = LF_E - D_E = 15 \text{ days} - 6 \text{ days}$$
$$= 9 \text{ days}$$

For activity D,

$$LF_D = LS_F = 15 \text{ days}$$
$$LS_D = LF_D - D_D = 15 \text{ days} - 3 \text{ days}$$
$$= 12 \text{ days}$$

For activity C, the LF value will be the lower LS value of its successors, activity E and activity G,

$$LF_C = LS_E = 9 \text{ days}$$
$$LS_C = LF_C - D_C = 9 \text{ days} - 6 \text{ days}$$
$$= 3 \text{ days}$$

For activity B, the LF value will be the lower LS value of its successors, activity D and activity E.

$$LF_B = LS_E = 9 \text{ days}$$
$$LS_B = LF_B - D_B = 9 \text{ days} - 5 \text{ days}$$
$$= 4 \text{ days}$$

For activity A, the LF value will be the lower LS value of its successors, activity B and activity C.

$$LF_A = LS_C = 3 \text{ days}$$
$$LS_A = LF_A - D_A = 3 \text{ days} - 3 \text{ days}$$
$$= 0 \text{ days}$$

Calculate the total float (TF) for each activity using the equation, $TF = LF - EF = LS - ES$. The TF is always zero for the first and last activities.

For activity A,

$$TF_A = LS_A - ES_A = 0 \text{ days} - 0 \text{ days}$$
$$= 0 \text{ days}$$

For activity B,

$$TF_B = LS_B - ES_B = 4 \text{ days} - 3 \text{ days}$$
$$= 1 \text{ day}$$

For activity C,

$$TF_C = LS_C - ES_C = 3 \text{ days} - 3 \text{ days}$$
$$= 0 \text{ days}$$

For activity D,

$$TF_D = LS_D - ES_D = 12 \text{ days} - 8 \text{ days}$$
$$= 4 \text{ days}$$

For activity E,

$$TF_E = LS_E - ES_E = 9 \text{ days} - 9 \text{ days}$$
$$= 0 \text{ days}$$

For activity F,

$$TF_F = LS_F - ES_F = 15 \text{ days} - 15 \text{ days}$$
$$= 0 \text{ days}$$

For activity G,

$$TF_G = LS_G - ES_G = 20 \text{ days} - 9 \text{ days}$$
$$= 11 \text{ days}$$

For activity H,

$$TF_H = LS_H - ES_H = 22 \text{ days} - 22 \text{ days}$$
$$= 0 \text{ days}$$

The TF for a project's critical path is always zero. Since A-C-E-F-H has a TF of zero, it is the critical path.

The CPM path is as shown.

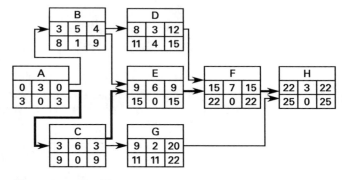

The answer is (D).

Why Other Options Are Wrong

(A) This incorrect option is not the critical path (activity G has a float that is not zero).

(B) This incorrect option is not the critical path (activities B and D have floats that are not zero).

(C) This incorrect option is not the critical path (activity B has a float that is not zero).

61. A *start-to-start* diagram shows that the start of B depends on the start of A, plus lag time.

A *start-to-finish diagram* (which is rarely used) shows that the finish of B depends on the start of A, plus lag time.

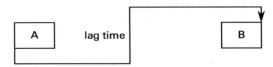

A *finish-to-start diagram* shows that the start of B depends on the finish of A, plus lag time.

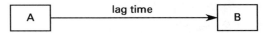

A *finish-to-finish diagram* shows that the finish of B depends on the finish of A, plus lag time.

The answer is (C).

Why Other Options Are Wrong

(A) This incorrect option assumed the diagram represented a start-to-finish relationship.

(B) This incorrect option assumed the diagram represented a finish-to-start relationship.

(D) This incorrect option assumed the diagram represented a finish-to-finish relationship.

62. *Fixed costs* (also known as *indirect costs*) are overhead costs that are not directly associated with specific project activities. Such costs include rent, utilities, and administrative staff. *Direct costs* (also known as *variable costs*) are costs that are directly associated with the completion of a project. Such costs include rental equipment, fuel, and labor.

The answer is (D).

Why Other Options Are Wrong

(A) This incorrect option represents a fixed cost.

(B) This incorrect option represents a fixed cost.

(C) This incorrect option represents a fixed cost.

63. Find the free float (FF) for activities B, D, E, and F.

The FF of activity B (with successors D and E) is

$$FF_B = ES_D - EF_B = 8 \text{ days} - 8 \text{ days}$$
$$= 0 \text{ days}$$

The FF of activity D (with successors E and H) is

$$FF_D = ES_E - EF_D = 9 \text{ days} - 11 \text{ days}$$
$$= -2 \text{ days}$$

The FF of activity E (with successor H) is

$$FF_E = ES_H - EF_E = 19 \text{ days} - 15 \text{ days}$$
$$= 4 \text{ days}$$

Because 1 day of lag time is needed from F to G, this must be subtracted from the FF. The FF of activity F (with successors G and H) is

$$FF_F = ES_G - EF_F - (F - S)$$
$$= 17 \text{ days} - 16 \text{ days} - 1 \text{ day}$$
$$= 0 \text{ days}$$

Activity E has the greatest FF of 4 days.

The answer is (C).

Why Other Options Are Wrong

(A) This incorrect option gives the answer as activity B, which does not have the greatest FF.

(B) This incorrect option gives the answer as activity D, which does not have the greatest FF.

(D) This incorrect option gives the answer as activity F, which does not have the greatest FF.

64. A project's crash time is equal to its earliest finish time. To determine the cost of the project completed under crash time, first find the cost per day for the project under normal time.

$$\frac{\$64{,}000}{25 \text{ days}} = \$2560/\text{day}$$

Then, determine the number of days, n, the project's duration can be reduced. Cost increases 10% per each day reduced.

$$\left(\left(2560 \ \frac{\$}{\text{day}} \right) (25 \text{ days} - n) \right) (1 + 0.1n) = \$72{,}000$$
$$(25 \text{ days} - n)(1 + 0.1n) = 28.125$$
$$-0.1n + 1.5n + 25 = 28.125$$
$$0.1n^2 - 1.5n + 3.125 = 0$$

$$n = \frac{-(-1.5) \pm \sqrt{(-1.5)^2 - (4)(0.1)(3.125)}}{(2)(0.1)}$$
$$= 2.5 \text{ days or } 12.5 \text{ days}$$

The optimal time is 2.5 days (2 days). Since the project can be reduced by 2 days, the earliest the project can be completed is in

$$25 \text{ days} - 2 \text{ days} = 23 \text{ days}$$

Therefore, under crash time, the cost of the project per day is

$$\left(2560 \ \frac{\$}{\text{day}} \right) (1 + 0.1n) = \left(2560 \ \frac{\$}{\text{day}} \right) (1 + (0.1)(2))$$
$$= \$3072/\text{day} \quad (\$3100/\text{day})$$

The answer is (C).

Why Other Options Are Wrong

(A) This incorrect option calculated the cost per day for the project completed under normal time.

(B) This incorrect option calculated the cost per day assuming that the project could be reduced by 1 day, not 2 days.

(D) This incorrect option calculated the cost per day assuming that the project could be reduced by 3 days, not 2 days.

65. A project's projected total final costs upon completion are known as the *estimate at completion* (EAC). The EAC can be found using the equation in option B, adding the project's *actual cost of work performed* (ACWP), which is the cost of the project up to this point, and its *estimate to complete* (ETC), which is the projected cost of the project from this point forward. That is, EAC = ACWP + ETC.

The *budget at completion* (BAC) is the total budgeted cost of the project. The *budgeted cost of work performed* (BCWP), is the budgeted cost of the project up to this point. The difference between BAC and BCWP is ETC. That is, ETC = BAC − BCWP.

The answer is (B).

Why Other Options Are Wrong

(A) This incorrect option represents the ETC, which is the estimated cost of the remaining work, not the total final cost. The equation for ETC is correct, however.

(C) This incorrect option represents the ETC, which is the estimated cost of the remaining work, not the total final cost. Also, the equation presented is incorrect because ETC is the difference of the BAC and the BCWP, not their sum.

(D) This incorrect option represents the EAC, which is the estimate at completion cost. However, the equation presented is incorrect because the EAC is the sum of the ACWP and the ETC, not their difference.

66. Calculate the most likely completion time of the project. The most likely completion time is the sum of the mean activity times.

$$\mu_{\text{critical path}} = 16 \text{ days} + 11 \text{ days} + 9 \text{ days} = 36 \text{ days}$$

The time a project will vary from its estimated completion time is the sum of the variances along the critical path. Variance, σ^2, is the square of the standard deviation, σ.

$$\sigma^2_{\text{critical path}} = (1.5 \text{ days})^2 + (0.7 \text{ days})^2 + (3.1 \text{ days})^2$$

$$= 12.35 \text{ days}^2$$

$$\sigma_{\text{critical path}} = \sqrt{12.35 \text{ days}^2}$$

$$= 3.51 \text{ days} \quad (4 \text{ days})$$

The completion date for the project is

$$36 \text{ days} + 4 \text{ days} = 40 \text{ days}$$

The answer is (C).

Why Other Options Are Wrong

(A) This incorrect option simply calculated the total mean activity times.

(B) This incorrect option neglected to square the activity standard deviations, thus yielding an incorrect complete time.

(D) This incorrect option calculated the square of the standard deviation, thus yielding an incorrect completion time.

67. Assign y as the profit, and x_1 and x_2 as the amount of product 1 and product 2, respectively. The profit is represented as

$$y_\$ = 140x_1 + 160x_2$$

The limiting amount of clay is represented as

$$20x_1 + 40x_2 \leq 280 \text{ tons}$$

The limiting number of hours the blending machine may be run by an operator is represented as

$$5x_1 + 5x_2 \leq 50 \text{ hr}$$

The amount of product 1 and product 2 that may be produced is limited by the curing vats' capacities. For product 1, $x_1 \leq 8$ tons. For product 2, $x_2 \leq 6$ tons. It is assumed that both product 1 and product 2 will be greater than or equal to zero.

Plot these constraints.

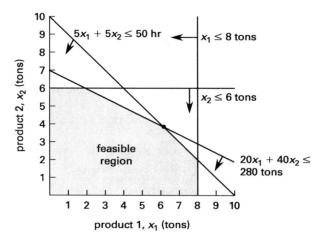

The shaded area represents the feasible region (where all constraints are satisfied). Points of intersection between constraints are the feasible points at which the profit will be evaluated. Find the point (x_1, x_2) at which the profit is maximized.

| | | profit at |
feasible points		feasible point
x_1 (tons)	x_2 (tons)	$y_\$ = 140x_1 + 160x_2$
0	6	$960
2	6	$1240
6	4	$1480
8	2	$1440
8	0	$1120

From the table, the maximum profit ($1480) occurs at $(x_1, x_2) = (6, 4)$. So, 6 tons of product 1 and 4 tons of product 2 should be produced.

The answer is (C).

Why Other Options Are Wrong

(A) This incorrect option assumed that because product 2 is more profitable per ton, all the resources should be used to manufacture it.

(B) This incorrect option assumed the point (2, 6) will maximize the profit.

(D) This incorrect option assumed the resources will be available for the curing vats' capacities.

68. The *crash time* is the shortest time in which an activity can be completed, whereas the *normal time* is the time in which an activity can be completed least expensively. For each activity on the critical path, the graph provided in the problem shows its normal time (the lowest point on that activity's cost curve) and its crash time (the leftmost point).

To keep a project's cost at a minimum, the duration of an activity is usually scheduled at its normal time.

When an activity's duration is made shorter than its normal time, a project's cost will increase. The slope to the left of any point shows how a project's cost will change if the activity's duration is reduced by one day.

Using the problem's graph, look at the slopes to the left of the normal times for the activities. The shallowest slope is in the curve for activity C. Start by reducing activity C from seven days to six days.

Next, look at the slopes to the left of the current durations for the activities (six days for activity C and the normal times for the other activities). The shallowest slope is in the curve for activity B, so reduce activity B from six days to five days. Repeat this examination one more time and reduce activity B from five days to four days.

Therefore, activity C should be shortened by one day, and activity B should be shortened by two days. The duration of activity C will be six days, and the duration of activity B will be four days.

The answer is (C).

Why Other Options Are Wrong

(A) This incorrect option is less cost-effective because activity C has a shallower slope than activity B on the first reduced day.

(B) This incorrect option is less cost-effective because the second reduced day from activity C has a steeper slope than the second reduced day from activity B.

(D) This incorrect option is less cost-effective because it includes activity D, which has a steeper slope than the second reduced day from activity B.

69. ACI 318 Sec. 7.4.1 states that when concrete is placed, reinforcements must be free from mud, oil, or other nonmetallic coatings that can decrease bonding. Thus, the mud must be cleaned (e.g., sand blasted) from the coated reinforcements.

ACI 318 Sec. 7.4.2 states that except for prestressing steel, steel reinforcement with rust, mill scale, or a combination of both is generally satisfactory. (The minimum dimensions, including height of deformations, and the weight of a test specimen wire-brushed by hand must comply with applicable ASTM specifications referenced in ACI 318 Sec. 3.5.) In this case, since the mill scale is minor, it should be acceptable according to ACI 318 Sec. 7.4.2.

The answer is (C).

Why Other Options Are Wrong

(A) This incorrect option unnecessarily rejects the entire shipment.

(B) This incorrect option unnecessarily rejects a portion of the shipment.

(D) This incorrect option neglects the requirements of ACI 318 Sec. 7.4.1.

70. The quantity of water, Q, flowing through the pipe is found using Darcy's law, $Q = KiA_{\text{gross}}$. K is the coefficient of permeability (in/sec), i is the hydraulic gradient (in/in), and A is the gross area (in^2). The hydraulic gradient, i, is the change in hydraulic head over a particular distance (in this case, the soil sample length, L).

$$i = \frac{\Delta H}{L} = \frac{10 \text{ in}}{20 \text{ in}}$$
$$= 0.5 \text{ in/in}$$

Therefore, the volume of water flowing through the pipe is

$$Q = KiA_{\text{gross}} = \left(5 \times 10^{-6} \, \frac{\text{in}}{\text{sec}}\right)\left(0.5 \, \frac{\text{in}}{\text{in}}\right)\left(\frac{\pi(2 \text{ in})^2}{4}\right)$$
$$= 7.85 \times 10^{-6} \text{ in}^3/\text{sec} \quad (8.0 \times 10^{-6} \text{ in}^3/\text{sec})$$

The answer is (C).

Why Other Options Are Wrong

(A) This incorrect option calculated the velocity of the water.

(B) This incorrect option is the permeability of the soil.

(D) This incorrect option miscalculated the area of the pipe by using the diameter of the soil sample as the radius.

71. The two pieces shown in the problem do not have grooves, so the T-joint must use a fillet weld to connect them. A fillet weld symbol is usually represented by a triangle placed on one or both sides of an arrow that points to the connecting point.

In this case, a fillet weld is present on each side of the connection. The top triangle of the welding symbol represents the size of the weld on the right side (opposite side, $1/4$ in), while the bottom triangle shows the size of the weld on the left side (arrow side, $5/16$ in).

The number to the right of the triangle identifies the weld length. Therefore, the weld on the arrow side (left side) is 6 in, and the weld on the right side is 4 in.

The answer is (A).

Why Other Options Are Wrong

(B) This incorrect option misinterprets the illustration.

(C) This incorrect option misinterprets the illustration.

(D) This incorrect option misinterprets the illustration.

72. There are typically three ways that anchor bolts fail: steel strength failure, concrete pull-out strength, or concrete breakout strength.

ACI 318 App. D defines the nominal strength of an anchor bolt by the equation: $N_{sa} = nA_{se}f_{uta}$ (ACI 318 Eq. D-3). N_{sa} is the achor nominal strength (lbf); n is the number of bolts; A_{se} is the effective cross-sectional area of anchor (in^2), and f_{uta} is the specified yield strength of anchor steel (lbf/in^2). If an anchor bolt's tensile force exceeds the calculated value of N_{sa}, the bolt will fail.

ACI 318 App. D defines anchor pull-out strength as "the strength corresponding to the anchoring device or a major component of the device sliding out from the concrete without breaking out a substantial portion of the surrounding concrete." It is found by the equation $N_p = 8A_{brg}f'_c$ (ACI 318 Eq. D-15). N_p is the pull-out strength in cracked concrete (lbf); and A_{brg} is the bearing area of the head of stud or anchor bolt (in^2). If the anchor's tensile force exceeds the calculated value of N_p, it will fail.

ACI 318 App. D defines concrete breakout by Eq. D-4 and Eq. D-7, where A_{NC} is the projected concrete failure area of a single anchor in tension (in^2); A_{NCO} is the projected concrete failure of a single anchor if not limited by edge distance or spacing (in^2); $\psi_{ed,N}$ is a factor used to modify anchor tensile strength based on proximity to edges of concrete member; $\psi_{c,N}$ is a factor used to modify tensile strength of anchors based on presence or absence of cracks in concrete; $\psi_{cp,N}$ is a factor used to modify tensile strength of post-installed anchors intended for use in uncracked concrete without supplementary reinforcement; N_b is the basic concrete breakout strength in tension of a single anchor in cracked concrete (lbf); k_c is the coefficient for basic concrete breakout strength in tension; and h_{ef} is the effective embedment depth of an anchor (in).

$$N_{cb} = \frac{A_{NC}}{A_{NCO}}\psi_{ed,N}\psi_{c,N}\psi_{cp,N}N_b \quad \text{[ACI 318 Eq. D-4]}$$

$$N_b = k_c\lambda\sqrt{f'_c}h_{ef}^{1.5} \quad \text{[ACI 318 Eq. D-7]}$$

If the anchor's concrete tensile force exceeds either the calculated value for N_{cb} or N_b, it will fail.

An anchor bolt experiencing a tensile force is shown.

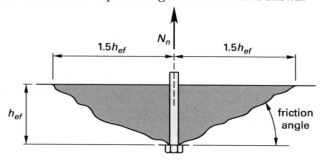

From this illustration and the ACI 318 App. D equations, the following will affect tensile strength.

- bolt length
- bolt diameter
- concrete edge distance
- concrete yield strength
- concrete internal friction angle

The magnitude of the tensile (pull-out) force does not affect tensile strength.

The answer is (B).

Why Other Options Are Wrong

(A) This incorrect option assumed the length of the anchor bolt will not affect tensile strength.

(C) This incorrect option assumed the compressive yield strength of concrete will not affect tensile strength.

(D) This incorrect option assumed the friction angle of concrete will not affect tensile strength.

73. *Quality control* evaluates the design, production, and construction quality of all aspects of a project. It involves testing, sampling, and inspecting products to find defects, as well as documenting, tracking, and trending. However, quality control is just one stage of the larger *quality assurance* process, which focuses on improving a product's production to minimize any issues that could lead to defects in that product. Quality assurance (not quality control) includes an independent assessment or audit stage and an acceptance stage. The results of the independent (i.e., third-party) audit are used to verify a project's own test results, and the acceptance stage consists of a series of final tests used to determine whether a product meets its required specifications.

The answer is (D).

Why Other Options Are Wrong

(A) This incorrect option assumed testing is not part of quality control.

(B) This incorrect option assumed inspecting is not part of quality control.

(C) This incorrect option assumed documenting is not part of quality control.

74. Calculate the cement weight and absolute volume. The cement weight is

$$W_{\text{cement}} = (6 \text{ sacks})\left(94 \ \frac{\text{lbf}}{\text{sacks}}\right)$$
$$= 564 \text{ lbf}$$

The cement volume is

$$V_{cement} = \frac{W_{cement}}{(SG)\gamma_{water}} = \frac{564 \text{ lbf}}{(3.5)\left(62.4 \frac{\text{lbf}}{\text{ft}^3}\right)}$$

$$= 2.58 \text{ ft}^3$$

Calculate the sand weight and volume. Using the cement to sand ratio (1:2.5), the sand weight is

$$W_{sand} = (564 \text{ lbf})(2.5) = 1410 \text{ lbf}$$

The sand volume is

$$V_{sand} = \frac{W_{sand}}{(SG)\gamma_{water}} = \frac{1410 \text{ lbf}}{(2.7)\left(62.4 \frac{\text{lbf}}{\text{ft}^3}\right)}$$

$$= 8.37 \text{ ft}^3$$

Calculate the aggregate weight and volume. Using cement to aggregate ratio (1:3.5), the aggregate weight is

$$W_{aggregate} = (564 \text{ lbf})(3.5) = 1974 \text{ lbf}$$

The aggregate volume is

$$V_{aggregate} = \frac{W_{aggregate}}{(SG)\gamma_{water}} = \frac{1974 \text{ lbf}}{(2.8)\left(62.4 \frac{\text{lbf}}{\text{ft}^3}\right)}$$

$$= 11.30 \text{ ft}^3$$

The total volume of cement, sand, and aggregate is

$$V_{total} = 2.58 \text{ ft}^3 + 8.37 \text{ ft}^3 + 11.30 \text{ ft}^3$$

$$= 22.25 \text{ ft}^3$$

Find the water volume per cubic yard of the mixture.

$$V_{water} = \left(27 \frac{\text{ft}^3}{\text{yd}^3}\right)(1 \text{ yd}^3) - 22.25 \text{ ft}^3$$

$$= 4.75 \text{ ft}^3$$

The weight of water in the cement mixture is

$$W_{water} = V_{water}\gamma_{water} = (4.75 \text{ ft}^3)\left(62.4 \frac{\text{lbf}}{\text{ft}^3}\right)$$

$$= 296 \text{ lbf}$$

Find the water-cement ratio.

$$\frac{W_{cement}}{W_{water}} = \frac{564 \text{ lbf}}{296 \text{ lbf}} = 1.91$$

The water-cement ratio is 1:1.91.

The answer is (A).

Why Other Options Are Wrong

(B) This incorrect option found the water-sand ratio, not the water-cement ratio.

(C) This incorrect option found the water-aggregate ratio, not the water-cement ratio.

(D) This incorrect option found the ratio of water to total concrete weight, not the water-cement ratio.

75. Find the cement's weight and volume. The total weight for the 5.5 sacks of cement is

$$W_{cement} = (5.5 \text{ sacks})\left(94 \frac{\text{lbf}}{\text{sack}}\right)$$

$$= 517 \text{ lbf}$$

The specific weight of water is 62.4 lbf/ft^3. The cement volume is

$$V_{cement} = \frac{W_{cement}}{(SG)\gamma_{water}} = \frac{517 \text{ lbf}}{(2.7)\left(62.4 \frac{\text{lbf}}{\text{ft}^3}\right)}$$

$$= 3.07 \text{ ft}^3$$

Find the water's weight and volume. The water-cement ratio is 1:2.5, so the water's weight is

$$W_{water} = \frac{517 \text{ lbf}}{2.5} = 206.8 \text{ lbf}$$

The specific gravity of water is 1.0. The water's volume is

$$V_{water} = \frac{W_{water}}{(SG)\gamma_{water}} = \frac{206.8 \text{ lbf}}{(1.0)\left(62.4 \frac{\text{lbf}}{\text{ft}^3}\right)}$$

$$= 3.31 \text{ ft}^3$$

Find the air's volume. The air is 5% of the mixture, so its volume is

$$V_{air} = \left(27 \frac{\text{ft}^3}{\text{yd}^3}\right)(1 \text{ yd}^3)(0.05) = 1.35 \text{ ft}^3$$

Calculate the volume and weight of the aggregates. The total volume of aggregates in one cubic yard of the mixture is

$$V_{total} = \left(27 \frac{\text{ft}^3}{\text{yd}^3}\right)(1 \text{ yd}^3) - 3.07 \text{ ft}^3 - 3.31 \text{ ft}^3 - 1.35 \text{ ft}^3$$

$$= 19.27 \text{ ft}^3$$

The mixture has 50% coarse aggregates and 50% fine aggregates by volume.

The volume of fine aggregates is

$$V_{fine} = (19.27 \text{ ft}^3)(0.5) = 9.64 \text{ ft}^3$$

The volume of coarse aggregates is

$$V_{coarse} = (19.27 \text{ ft}^3)(0.5) = 9.64 \text{ ft}^3$$

The weight of fine aggregates is

$$W_{fine} = (9.64 \text{ ft}^3)(2.5)\left(62.4 \frac{\text{lbf}}{\text{ft}^3}\right)$$
$$= 1503.8 \text{ lbf}$$

The weight of coarse aggregates is

$$W_{coarse} = (9.64 \text{ ft}^3)(3.0)\left(62.4 \frac{\text{lbf}}{\text{ft}^3}\right)$$
$$= 1804.61 \text{ lbf}$$

The fine aggregates have a moisture content of 5%. Therefore, the weight of water in the fine aggregates is

$$W_{water,fine} = (1503.8 \text{ lbf})(0.05)$$
$$= 75.19 \text{ lbf}$$

The coarse aggregates have a moisture content of 2%. Therefore, the weight of water in the coarse aggregates is

$$W_{water,coarse} = (1805 \text{ lbf})(0.02)$$
$$= 36.1 \text{ lbf}$$

The total moisture weight in the aggregates is

$$W_{total} = W_{water,fine} + W_{water,coarse} = 75.19 \text{ lbf} + 36.1 \text{ lbf}$$
$$= 111.3 \text{ lbf}$$

The weight of water that must be added to the mix is

$$206.8 \text{ lbf} - 111.3 \text{ lbf} = 95.5 \text{ lbf} \quad (95 \text{ lbf})$$

The answer is (B).

Why Other Options Are Wrong

(A) This incorrect option did not include the volume of air in the concrete, which resulted in a larger aggregate volume.

(C) This incorrect option mistook the water content for the water weight to total aggregate weight ratio.

(D) This incorrect option did not subtract the aggregate water weight from the total water before determining the amount of water to be added.

76. Find the lateral wind velocity pressure, q. The wind velocity is 110 mph; however, for temporary structures used less than six weeks, a reduction factor of 0.75 must be applied (ASCE 37 Sec. 6.2.1). Since the formwork will only stand for two weeks, apply the 0.75 reduction factor to the wind velocity. Also, for one-minute average wind speeds, a factor of 1.25 must be applied to the wind velocity (ASCE 37 Sec. C6.2.1.2). ASCE 37 points to

ASCE 7 Eq. 6-15 to determine the wind velocity pressure. (This equation is not dimensionally consistent.)

$$q_{wind,psf} = 0.00256 K_z K_{zt} K_d v^2 I$$
$$= (0.00256)(1.0)(1.0)(1.0)$$
$$\times \left(\left(110 \frac{\text{mi}}{\text{hr}}\right)(0.75)(1.25)\right)^2 (1.0)$$
$$= 27.23 \text{ lbf/ft}^2$$

The importance factor, I, is 1.0 for all environmental loads, per ASCE 37 Sec. 6.1.

The total wind force, F_{wind}, acting on the 2 ft × 12 ft tall column is

$$F_{wind} = q_{wind,psf} w_{formwork} h_{formwork}$$
$$= \left(27.23 \frac{\text{lbf}}{\text{ft}^2}\right)(2 \text{ ft})(12 \text{ ft})$$
$$= 653.52 \text{ lbf}$$

Calculate the force in brace B.

Take the moment around the bottom of the column, with a clockwise moment from the wind and a counterclockwise from the vertical reaction of the pin. Since the concrete column and form are in equilibrium, the moment from the wind load and the moment from the vertical reaction of the pin must cancel out.

$$M_{bottom} = (653.52 \text{ lbf})(6 \text{ ft}) - P_y(8 \text{ ft}) = 0$$
$$P_y = \frac{(653.52 \text{ lbf})(6 \text{ ft})}{8 \text{ ft}}$$
$$= 490.14 \text{ lbf}$$

The horizontal pin reaction, P_x, has the same magnitude as the wind force. So, P_x is 653.52 lbf.

From static equilibrium, the forces at the pin must cancel out. Therefore, the vertical force from brace A must equal the vertical reaction of the pin.

$$F_A \sin\theta = P_y = 490.14 \text{ lbf}$$
$$F_A = 693.16 \text{ lbf} \quad \text{[compression]}$$

Using left as negative, right as positive, the horizontal force from brace B is

$$F_B + F_A \cos\theta - P_x = 0$$
$$F_B + 490.14 \text{ lbf} - 653.52 \text{ lbf} = 0$$
$$F_B = 163.38 \text{ lbf}$$
$$(160 \text{ lbf}) \quad \text{[tension]}$$

The answer is (B).

Why Other Options Are Wrong

(A) This incorrect option did not account for the wind suction factor.

(C) This incorrect option did not multiply the wind force by the temporary structure reduction factor before calculating the lateral wind pressure.

(D) This incorrect option found the force in brace A, not brace B.

77. (a) Formwork must be able to withstand hydraulic loading from concrete during curing. Equations for lateral form pressures are provided in ASCE 37 Sec. 4.7. For walls placed at a rate of less than 7 ft/hr, ASCE 37 Eq. 4-3 applies. (Note that ASCE 37 uses C_c for lateral force pressure, whereas p_{max} is used in this solution.) To comply with ASCE 37 Sec. 4.7, p_{max} is limited to a maximum of 2000 lbf/ft^2 and a minimum of 600 lbf/ft^2. (This equation is not dimensionally consistent.)

$$p_{max,psf} = 150 \ \frac{\text{lbf}}{\text{ft}^3} + 9000 \left(\frac{R_{\text{ft/hr}}}{T_{\circ F}} \right)$$
$$= 150 \ \frac{\text{lbf}}{\text{ft}^3} + (9000) \left(\frac{5 \ \frac{\text{ft}}{\text{hr}}}{60 \,^\circ \text{F}} \right)$$
$$= 900 \ \text{lbf/ft}^2 \quad [\text{OK}]$$

However, the wet concrete acts as a fluid, and the lateral pressure of concrete on the form increases linearly with depth until it reaches the maximum lateral pressure of 900 lbf/ft^2. Calculate the depth at which this maximum pressure is reached by rearranging ASCE 37 Eq. 4-1.

$$h = \frac{p_{max}}{w} = \frac{900 \ \frac{\text{lbf}}{\text{ft}^2}}{150 \ \frac{\text{lbf}}{\text{ft}^3}}$$
$$= 6 \ \text{ft}$$

(b) According to ASCE 37 Sec. 4.7.1.1, the lateral pressure below 6 ft is also 900 lbf/ft^2 since it is the maximum value. Therefore, from the bottom of the formwork, a maximum lateral concrete pressure of 900 lbf/ft^2 will occur up to a depth of 6 ft, and decreases linearly until the pressure drops to zero at the top.

The answer is (B).

Why Other Options Are Wrong

(A) This incorrect option assumed the concrete unit weight to be identical to the maximum lateral concrete pressure.

(C) This incorrect option assumed that maximum pressure occurs at the bottom of the formwork instead of calculating the correct depth.

(D) This incorrect option calculated the maximum lateral concrete pressure by multiplying the concrete unit weight by the height of the formwork instead of using ASCE 37 Eq. 4-3.

78. Use ASCE 37 Eq. 4-3 to find the maximum lateral concrete pressure. To comply with ASCE 37 Sec. 4.7, p_{max} must be greater than 600 lbf/ft^2 and less than 2000 lbf/ft^2. (This equation is not dimensionally consistent.)

$$p_{max,psf} = 150 \ \frac{\text{lbf}}{\text{ft}^3} + 9000 \left(\frac{R_{\text{ft/hr}}}{T_{\circ F}} \right)$$
$$= 150 \ \frac{\text{lbf}}{\text{ft}^3} + (9000) \left(\frac{5 \ \frac{\text{ft}}{\text{hr}}}{75 \,^\circ \text{F}} \right)$$
$$= 750 \ \text{lbf/ft}^2 \quad [\text{OK}]$$

Find the uniform load on a 12 in wide strip of horizontal sheathing.

$$w = \left(750 \ \frac{\text{lbf}}{\text{ft}^2} \right) \left(\frac{12 \ \text{in}}{12 \ \frac{\text{in}}{\text{ft}}} \right) = 750 \ \text{lbf/ft}$$

ACI SP-4 Table 7-1 provides formulas for calculating the span length that the sheathing can expand without the support of studs. (These formulas are not dimensionally consistent.) Furthermore, variables used in ACI SP-4 Table 7-1 differ slightly from those used in Table 4-2 and Table 4-3. From the problem statement and ACI Table 7-1, the bending stress is $F_b = F_b' = 1545$ lbf/in^2. The section modulus is $KS = S = 0.664$ in^3. Therefore, the maximum support spacing for the bending is

$$l_{in} = 10.95 \sqrt{\frac{F_{b,\text{lbf/in}^2}' S_{\text{in}^3}}{w_{\text{lbf/ft}}}}$$
$$= 10.95 \sqrt{\frac{\left(1545 \ \frac{\text{lbf}}{\text{in}^2} \right)(0.664 \ \text{in}^3)}{750 \ \frac{\text{lbf}}{\text{ft}}}}$$
$$= 12.81 \ \text{in} \quad [\text{controls}]$$

The rolling shear stress is $F_v = F_s' = 57$ lbf/in^2, and the rolling shear constant, Ib/Q, is 8.882 in^2. From ACI SP-4 Table 7-1, the maximum support spacing that will accommodate rolling shear in the sheathing is

$$l_{in} = \left(\frac{20 F_{s,\text{lbf/in}^2}'}{w_{\text{lbf/ft}}} \right) \left(\frac{Ib}{Q} \right)_{\text{in}^2} + 1.5$$
$$= \left(\frac{(20) \left(57 \ \frac{\text{lbf}}{\text{in}^2} \right)}{750 \ \frac{\text{lbf}}{\text{ft}}} \right) (8.882 \ \text{in}^2) + 1.5$$
$$= 15.006 \ \text{in}$$

Find the maximum support spacing that will accommodate deflection in the sheathing. Using ACI SP-4 Table 7-1 for a maximum deflection of 1/360, calculate the support spacing. The modulus of elasticity is 1,500,000 lbf/in². The moment of inertia, I, is obtained from ACI SP-4 Table 4-3 as 0.423 in⁴.

$$l_{in} = 1.69 \sqrt[3]{\frac{E'_{lbf/in^2} I_{in^4}}{w_{lbf/ft}}}$$

$$= 1.69 \sqrt[3]{\frac{\left(1,500,000 \frac{lbf}{in^2}\right)(0.423 \ in^4)}{750 \frac{lbf}{ft}}}$$

$$= 15.98 \ in$$

The bending spacing of 12.81 in will control the stud spacing. In design, it is common to round the maximum spacing down to a number that is multiple of 3 in. Therefore, a spacing of 12 in is appropriate.

The answer is (B).

Why Other Options Are Wrong

(A) This incorrect option chose a spacing less than that required.

(C) This incorrect option chose the highest maximum spacing rather than the lowest maximum spacing.

(D) This incorrect option chose the highest maximum spacing rather than the lowest maximum spacing. Furthermore, it rounded up to the next quarter inch.

79. Use ASCE 37 Eq. 4-3 to find the maximum lateral concrete pressure, p_{max}. To comply with ASCE 37 Sec. 4.7, p_{max} must be greater than 600 lbf/ft² and less than 2000 lbf/ft². (This equation is not dimensionally consistent.)

$$p_{max,psf} = 150 \frac{lbf}{ft^3} + 9000\left(\frac{R_{ft/hr}}{T_{°F}}\right)$$

$$= 150 \frac{lbf}{ft^3} + (9000)\left(\frac{6 \frac{ft}{hr}}{45°F}\right)$$

$$= 1350 \ lbf/ft^2 \quad [OK]$$

The tributary area is dependent on the spacing of the wales and form ties.

$$A_{trib} = l_{wale} l_{tie} = \frac{(18 \ in)(24 \ in)}{\left(12 \frac{in}{ft}\right)^2}$$

$$= 3 \ ft^2$$

Once the tributary area is calculated, the load on the tie can be found.

$$P = p_{max} A_{trib} = \left(1350 \frac{lbf}{ft^2}\right)(3 \ ft^2)$$

$$= 4050 \ lbf$$

The minimum cross-sectional area that will support this load is

$$A_{min} = \frac{P}{\phi F_y} = \frac{4050 \ lbf}{(0.9)\left(24 \frac{kips}{in^2}\right)\left(1000 \frac{lbf}{kip}\right)}$$

$$= 0.188 \ in^2$$

The diameter that corresponds to this cross-sectional area is

$$A_{cross} = \frac{\pi d^2}{4}$$

$$d = \sqrt{\frac{4 A_{cross}}{\pi}} = \sqrt{\frac{(4)(0.188 \ in^2)}{\pi}}$$

$$= 0.489 \ in$$

The commercially available form tie that meets the minimum allowable diameter is ¹/₂ in (0.50 in).

The answer is (B).

Why Other Options Are Wrong

(A) This incorrect option used a diameter smaller than was necessary.

(C) This incorrect option used a diameter greater than was necessary.

(D) This incorrect option used a diameter greater than was necessary.

80. Find the length of the strut, l, using the Pythagorean theorem. h' is the height of the top of the strut and l' is the horizontal distance from the form to the bottom of the strut.

$$l^2 = h'^2 + l'^2$$

$$l = \sqrt{h'^2 + l'^2} = \sqrt{(4 \ ft)^2 + (3 \ ft)^2}$$

$$= 5 \ ft$$

Derive the axial load, p', per foot on the brace. H is the lateral (wind) load at the top of the form, and h is the height of the form.

$$\cos \theta = \frac{l'}{l}$$

$$(p' \cos \theta) h' = Hh$$

$$p'\left(\frac{l'}{l}\right) h' = Hh$$

Therefore, the axial load is

$$p' = \frac{Hhl}{h'l'} = \frac{\left(80 \frac{lbf}{ft}\right)(5 \ ft)(5 \ ft)}{(4 \ ft)(3 \ ft)}$$

$$= 166.7 \ lbf/ft$$

The slenderness ratio of the strut should be kept under 50 so that no additional bracing is needed.

$$\frac{l}{d} = \frac{(5\text{ ft})\left(12\ \dfrac{\text{in}}{\text{ft}}\right)}{1.5\text{ in}} = 40 \quad [< 50,\text{ so OK}]$$

Find the allowable compressive stress on the brace for buckling.

$$F'_c = \frac{0.3E'_{\text{lbf/in}^2}}{\left(\dfrac{l_{\text{in}}}{d_{\text{in}}}\right)^2} = \frac{(0.3)\left(1{,}200{,}000\ \dfrac{\text{lbf}}{\text{in}^2}\right)}{\left(\dfrac{(5\text{ ft})\left(12\ \dfrac{\text{in}}{\text{ft}}\right)}{1.5\text{ in}}\right)^2}$$

$$= 225\text{ lbf/in}^2 < 800\text{ lbf/in}^2 \quad [\text{controls}]$$

The maximum axial load that the strut can withstand is

$$p = \left(225\ \frac{\text{lbf}}{\text{in}^2}\right)(1.5\text{ in})(3.5\text{ in})$$

$$= 1181\text{ lbf}$$

The axial pressure from the wind was previously calculated as 166.7 lbf/ft. Therefore, the maximum strut spacing is

$$s = \frac{p}{p'} = \frac{1181\text{ lbf}}{166.7\ \dfrac{\text{lbf}}{\text{ft}}}$$

$$= 7.08\text{ ft} \quad (7.0\text{ ft})$$

The answer is (A).

Why Other Options Are Wrong

(B) This incorrect option used actual dimensions of the brace instead of nominal dimensions.

(C) This incorrect option did not convert the lateral wind pressure to axial pressure.

(D) This incorrect option assumed the strut's allowable compressive stress was the same as the maximum compressive stress.

81. ACI 347 defines *shores* (not reshores) as "vertical or inclined support members designed to carry the weight of the formwork, concrete, and construction loads above." *Reshores* are defined as "shores placed snugly under a stripped concrete slab or other structural member after the original forms and shores have been removed..." Before reshores may be installed, the concrete slab or structural member must be able to deflect and support its own weight, as well as supplied existing loads. However, when construction of the upper floors finishes, the reshoring can support the concrete and formwork weight that is transferred from the shoring of upper floors to the concrete that is immediately above the reshoring. Since the reshoring is snugly fitted under the concrete slab, it will take any additional weight that is transferred from floors above immediate support level.

The answer is (A).

Why Other Options Are Wrong

(B) This incorrect option is a correct statement about reshores—they must be placed snugly under a stripped concrete slab or other structural member.

(C) This incorrect option is a correct statement about reshores—they are installed only after a concrete slab is able to support its own weight.

(D) This incorrect option is a correct statement about reshores—they are placed after shores and original forms are removed.

82. *Fly ash* is an admixture that is added to concrete to modify the concrete's performance. It is often added to cement because it increases the cement's strength and durability, while *decreasing* its permeability. As cement sets, calcium silicate hydrate and calcium hydroxide are formed. The fly ash reacts with the calcium hydroxide and increases binding.

The answer is (D).

Why Other Options Are Wrong

(A) This incorrect option represented a potential result of adding fly ash to a concrete mixture—binding is increased.

(B) This incorrect option represented a potential result of adding fly ash to a concrete mixture—strength is increased.

(C) This incorrect option represented a potential result of adding fly ash to a concrete mixture—durability is increased.

83. The flexural moment capacity of a wide flange steel beam is listed under "Part 3—Design of Flexural Members" in the *AISC Manual*. When a section is compacted, such as with a W24 × 68, the beam has full capacity when the unbraced length is smaller than L_p (the lateral unbraced length of the limit state of yielding).

When the unbraced length is in between L_p and L_b (the lateral unbraced length of the limit state of inelastic-torsional buckling), the moment capacity of the steel beam is subject to inelastic lateral-torsional buckling.

When the unbraced length is greater than L_b, the beam is subject to elastic lateral-torsional buckling. The moment capacity with unbraced length is presented in the *AISC Manual* Table 3-10 for W shapes. If the maximum moment in LRFD is 600 ft-kips, the beam does not have enough capacity with an unbraced length of 16 ft, and it must be laterally braced at some point to increase the moment capacity.

According to the *AISC Manual* Table 3-10, the maximum unbraced length is approximately 9.5 ft for a moment capacity of 600 ft-kips (LRFD).

The answer is (D).

Why Other Options Are Wrong

(A) This incorrect option stated that the beam is adequate to carry 600 ft-kips, but it is not adequate to do so.

(B) This incorrect option gave a span length of 13.5 ft, which would provide inadequate support for the load.

(C) This incorrect option gave a span length of 11.5 ft, which would provide inadequate support for the load.

84. Calculate the area of the sign.

$$A_{sign} = bh = (5 \text{ ft})(3 \text{ ft})$$
$$= 15 \text{ ft}^2$$

The total wind load, w, on the sign is

$$w_{total} = w_{wind}A_{sign} = \left(40 \frac{\text{lbf}}{\text{ft}^2}\right)(15 \text{ ft}^2)$$
$$= 600 \text{ lbf}$$

Assume that the wind load acts on the center of the sign. The moment arm is the distance from the centroid of the sign to the base of the sign plus the pole length. The moment that the wind load will generate is

$$M = (600 \text{ lbf})(1.5 \text{ ft} + 6 \text{ ft})$$
$$= 4500 \text{ ft-lbf}$$

The moment produced from the wind is resisted by the coupled moment from the anchor groups. The space between the anchor groups is 10 in. From statics, the tensile force, T_f, in the front anchor group and the compression force in the back anchor group will be the same.

$$M = T_f d$$
$$T_f = \frac{M}{d} = \frac{(4500 \text{ ft-lbf})\left(12 \frac{\text{in}}{\text{ft}}\right)}{10 \text{ in}}$$
$$= 5400 \text{ lbf}$$

Each group consists of two bolts. Accounting for the factor of safety, FS, the tensile capacity, T_c, of one bolt will be

$$T_c = \frac{T_f(\text{FS})}{2} = \frac{(5400 \text{ lbf})(3)}{2}$$
$$= 8100 \text{ lbf}$$

The answer is (C).

Why Other Options Are Wrong

(A) This incorrect option did not consider the factor of safety.

(B) This incorrect option calculated the moment generated by wind with a moment arm the length of the pole.

(D) This incorrect option gave the moment capacity as the total capacity of two bolts on either the front or back side of the base.

85. Find the area that is supported by each soldier pile.

$$A = (6 \text{ ft})(8 \text{ ft})$$
$$= 48 \text{ ft}^2$$

Calculate the dynamic force, F_D, of the water flow. g_c is the gravitational constant, which is 32.2 lbm-ft/lbf-sec^2.

$$F_D = \frac{C_D A \rho v^2}{2g_c} = \frac{(1.0)(48 \text{ ft}^2)\left(62.4 \frac{\text{lbm}}{\text{ft}^3}\right)\left(3 \frac{\text{ft}}{\text{sec}}\right)^2}{(2)\left(32.2 \frac{\text{lbm-ft}}{\text{lbf-sec}^2}\right)}$$
$$= 418.58 \text{ lbf}$$

The river is 8 ft deep and the water flow has a uniform depth, so the water's force will occur at the middle of the sheetpiles. This makes the moment arm 4 ft. Therefore, the moment at the riverbed is

$$M = F_D(\text{moment arm}) = (418.58 \text{ lbf})(4 \text{ ft})$$
$$= 1674.32 \text{ ft-lbf} \quad (1700 \text{ ft-lbf})$$

The answer is (A).

Why Other Options Are Wrong

(B) This incorrect option calculated the static water force.

(C) This incorrect option calculated the moment created by the static water force.

(D) This incorrect option calculated the moment created by the static water force using the full water depth of 8 ft as the moment arm.

86. The minimum area of tension, A_s, per ACI 318 Sec. 10.5.1, is limited by

$$A_{s,min} \geq \begin{cases} \dfrac{3\sqrt{f_c'}b_w d}{f_y} \\ \dfrac{200 b_w d}{f_y} \end{cases}$$

f'_c is the compressive strength, f_y is the tension yield strength, b_w is the footing width, and d is the effective beam depth. Therefore,

$$A_{s,\min} \geq \begin{cases} \dfrac{3\sqrt{f'_c}\,b_w d}{f_y} = \dfrac{3\sqrt{4000\ \dfrac{\text{lbf}}{\text{in}^2}}(10\text{ ft})\left(12\ \dfrac{\text{in}}{\text{ft}}\right)(21\text{ in})}{60{,}000\ \dfrac{\text{lbf}}{\text{in}^2}} \\ \qquad\qquad = 7.97\text{ in}^2 \\[2ex] \dfrac{200 b_w d}{f_y} = \dfrac{(200)(10\text{ ft})\left(12\ \dfrac{\text{in}}{\text{ft}}\right)(21\text{ in})}{60{,}000\ \dfrac{\text{lbf}}{\text{in}^2}} \\ \qquad\qquad = 8.40\text{ in}^2 \quad (8.5\text{ in}^2) \quad \text{[controls]} \end{cases}$$

The answer is (B).

Why Other Options Are Wrong

(A) This incorrect option does not take into account the fact that the reinforcement should not be less than $200 b_w d/f_y$.

(C) This incorrect option did not subtract 3 in from the total depth and so used an incorrect value for the effective beam depth.

(D) This incorrect option represents the area of reinforcement in both directions, rather than the reinforcement each way.

87. OSHA Std. 1926.451 gives scaffolding regulations. According to OSHA Subpart L (1926.451(g)(1)), guardrails must be provided on scaffolds that are more than 10 ft above a lower level, so option A is incorrect.

The 6 ft threshold for fall protection described in OSHA Subpart M applies to other construction walking/working surfaces, but not to scaffolds. Scaffolds are given a higher threshold because, as they are temporary structures designed to be erected and dismantled easily, they are not suited for the use of guardrails and other fall protection measures at the time that the lower level is being constructed.

The answer is (A).

Why Other Options Are Wrong

(B) This incorrect option represented a true statement per OSHA scaffolding regulations.

(C) This incorrect option represented a true statement per OSHA scaffolding regulations.

(D) This incorrect option represented a true statement per OSHA scaffolding regulations.

88. According to OSHA Std. 1926.652, all excavations deeper than 5 ft (except those in stable rock) must be protected from cave-in and collapse.

OSHA Std. 1926.652 App. B gives the following maximum allowable slopes for excavations less than 20 ft deep.

soil or rock type	maximum allowable slopes (deg)
stable rock	90°
type A	53° long-term;
	63° short-term
type B	45°
type C	34°

Soil types are defined in OSHA Std. 1926 Subpart P App. A as follows.

Type A soils are cohesive soils with an unconfined compressive strength of 1.5 tons per square foot or greater. Examples include clay, silty clay, sandy clay, clay loam, and, in some cases, silty clay loam and sandy clay loam.

Type B soils are cohesive soils with an unconfined compressive strength greater than 0.5 tons per square foot, but less than 1.5 tons per square foot. Examples include angular gravel, silt, silt loam, sandy loam, and, in some cases, silty clay loam and sandy clay loam.

Type C soils are cohesive soils with an unconfined compressive strength of 0.5 tons per square foot or less. Examples include granular soils such as gravel, sand, and loamy sand.

The problem states that the excavation will take place in sand, which is a type C soil. Therefore, the maximum allowable slope is 34°.

The answer is (D).

Why Other Options Are Wrong

(A) This incorrect option gave the maximum allowable slope for stable rock.

(B) This incorrect option gave the maximum allowable slope for type A soil (long-term).

(C) This incorrect option gave the maximum allowable slope for type B soil.

89. OSHA Std. 1904 Subpart B and Subpart C specify the scope and criteria of record keeping and forms for fatalities and injuries.

OSHA Std. 1904.29, item (a) states that employers must use OSHA forms 300 ("Log of Work-Related Injuries and Illnesses"), 300A ("Summary of Work-Related Injuries and Illnesses"), or 301 ("Injury and Illness Incident Report") to record workplace injuries and illnesses.

OSHA Std. 1904.1, item (a) states that if an employer has ten or fewer employees at all times during a calendar year, OSHA safety records do not need to be kept.

OSHA Std. 1904.5 covers accidents that are not considered work-related. OSHA Table 1904.5(b)(2) specifies that accidents are not considered work-related if "the injury or illness is caused by a motor vehicle accident and occurs on a company parking lot or company access road while the employee is commuting to or from work."

OSHA Std. 1904.39, item (a) states that within 8 hr after the death of any employee from a work-related incident or the in-patient hospitalization of three or more employees from a work-related incident, an employer must orally report the incident(s) to OSHA offices via phone or in person.

The answer is (C).

Why Other Options Are Wrong

(A) This incorrect option represented a true statement per OSHA regulations.

(B) This incorrect option represented a true statement per OSHA regulations.

(D) This incorrect option represented a true statement per OSHA regulations.

90. OSHA Std. 3000-08R, *Employer Rights and Responsibilities Following an OSHA Inspection*, provides guidance on employer rights and responsibilities after a workplace has been inspected by an OSHA compliance safety and health officer. If a violation exists, OSHA will issue the employer a citation and notification of penalty. There are six types of violations. (All definitions are from OSHA Std. 3000-08R.)

Repeated violation—given if a employer has been cited previously for a substantially similar condition and the citation has become a final order of the Occupational Safety and Health Review Commission. A citation is currently viewed as a repeated violation if it occurs within five years either from the date that the earlier citation becomes a final order or from the final abatement date, whichever is later. Repeated violations can bring a civil penalty of up to $70,000 for each violation.

Willful violation—given when an employer knew a hazardous condition existed, but made no reasonable effort to eliminate it, and in which the hazardous condition violated a standard, regulation, or the *OSH Act*. Penalties range from $5000 to $70,000 per willful violation.

Serious violation—given when the workplace hazard could cause injury or illness that would most likely result in death or serious physical harm, unless the employer did not know or could not have known of the violation. A penalty of up to $7000 for each violation may be imposed.

Failure to abate—given when the employer has not corrected a violation for which OSHA has issued a citation and the abatement date has passed or is covered under a settlement agreement. A failure to abate also exists when the employer has not complied with interim measures involved in a long-term abatement within the time given. A penalty of up to $7000 per day may be imposed for each violation.

Other-than-serious—given when the most serious injury or illness that would be likely to result from a hazardous condition cannot reasonably be predicted to cause death or serious physical harm to exposed employees, but does have a direct and immediate relationship to their safety and health. A penalty of up to $7000 may be imposed per violation.

De Minimis—given when a violation has no direct or immediate relationship to safety or health. This violation does not result in a citation or penalty.

Of these violations, only a failure to abate citation is given when an employer has not corrected for a previously issued OSHA violation.

The answer is (D).

Why Other Options Are Wrong

(A) This incorrect option chose the wrong violation type.

(B) This incorrect option chose the wrong violation type.

(C) This incorrect option chose the wrong violation type.

91. The *experience modification rate* (EMR) is a number used by insurance companies to assess the probability of future insurance claims from a company based on its previous claims. A company's EMR takes into account the nature of the work the company does, the average losses in insurance claims in the company's industry, and the insurance claims of the company over a recent three-year period.

A company whose insurance claims during this three-year period have been average for its industry will have an EMR of 1.0. A company with an EMR greater than 1.0 has had a higher-than-average rate of workers' compensation claims, which indicates a worse-than-average safety record, and this company will pay a higher-than-average workers' compensation insurance premium until a new EMR is assigned. On the other hand, a company with an EMR less than 1.0 has a better-than-average record and will pay a below-average workers' compensation premium until a new EMR is assigned.

The answer is (B).

Why Other Options Are Wrong

(A) This incorrect option assumed that a rating less than 1.0 did not indicate a better-than-average record.

(C) This incorrect option assumed the EMR is related to how long the company has been with the same insurer.

(D) This incorrect option assumed the EMR is related to how long the company has been with the same insurer.

92. The injury and illness incidence rate is found using the formula

$$\text{incidence rate} = \frac{(200{,}000 \text{ hr})\left(\begin{array}{c}\text{total number of}\\\text{injuries and illnesses}\end{array}\right)}{\text{hours worked by all employees}}$$

200,000 hr is the number of hours that 100 full-time employees will work in a calendar year, assuming a typical holiday schedule. The total number of injuries and illnesses is given as 12. The total number of hours worked by 800 employees working 40 hours a week for 50 weeks is

$$(800)\left(40 \ \frac{\text{hr}}{\text{wk}}\right)\left(50 \ \frac{\text{wk}}{\text{yr}}\right) = 1{,}600{,}000 \ \text{hr/yr}$$

Therefore, the injury and illness incidence rate for the company last year is

$$\begin{aligned}\text{incidence rate} &= \frac{(200{,}000 \text{ hr})\left(\begin{array}{c}\text{total number of}\\\text{injuries and illnesses}\end{array}\right)}{\text{hours worked by all employees}}\\[2mm] &= \frac{(200{,}000 \text{ hr})(12 \text{ injuries})}{1{,}600{,}000 \ \dfrac{\text{hr}}{\text{yr}}}\\[2mm] &= 1.5 \ \text{injuries/yr}\end{aligned}$$

The answer is (C).

Why Other Options Are Wrong

(A) This incorrect option was obtained by dividing the number of incidents by the number of employees.

(B) This incorrect option was obtained by assuming 52 working weeks in a year rather than 50.

(D) This incorrect option simply reported the number of incidents in the past year.

93. Groundwater control methods are generally broken into two categories: 1) groundwater exclusion, and 2) groundwater pumping.

Groundwater exclusion usually involves the use of walls to keep groundwater out of a confined area. Water is not removed from the ground. Typical methods in this category include the use of slurry walls, steel sheet-piles, and freeze walls. Slurry walls sometimes still allow some water to seep through, so a sump pump inside the confined area may also be needed.

Groundwater pumping is a straightforward method of keeping water out of a construction site. Sump pumping is applied when water depth is relatively shallow (less than 10 ft deep) with high soil permeability. Well points are used with medium soil permeability, and the depths of the wells needed vary depending on the depth of the groundwater level. Ejector wells are used when soil

permeability is low, and they can be applied to various depths underground. A vacuum pump is necessary when the permeability coefficient is lower than 2×10^{-3} in/sec.

The answer is (B).

Why Other Options Are Wrong

(A) This incorrect option assumed sump pumps are used for groundwater exclusion, when they are actually used for groundwater pumping.

(C) This incorrect option assumed all slurry pumps, well points, and ejector wells are used for groundwater exclusion, when they are actually used for groundwater pumping.

(D) This incorrect option assumed ejector wells are used for groundwater exclusion, when they are actually used for groundwater pumping.

94. The illustration shown represents the pattern of drilling and blasting. The smaller circles represent the drilled holes to place the explosives, and the bigger circles are the range of rock that is blasted by the explosives.

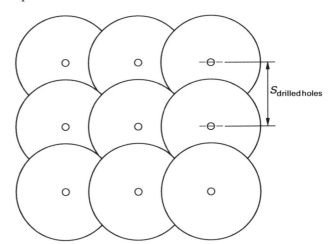

The volume of rock produced per hole, V_{hole}, is divided by the original hole depth, d_{hole}, to give the volume of rock per foot of hole drilled.

$$V_{\text{per ft of hole}} = \frac{V_{\text{hole}}}{d_{\text{hole}}}$$

Because of the pattern of the explosives, the volume of rock produced by each hole is estimated as a regular parallelepiped volume 8 ft on a side and 22 ft deep.

$$\begin{aligned}V_{\text{hole}} &= \frac{(8 \text{ ft})^2(22 \text{ ft})}{27 \ \dfrac{\text{ft}^3}{\text{yd}^3}}\\[2mm] &= 52.1 \ \text{yd}^3\end{aligned}$$

Therefore, the volume of rock produced per foot of drilling is

$$V_{\text{per ft of hole}} = \frac{V_{\text{hole}}}{d_{\text{hole}}} = \frac{52.1 \text{ yd}^3}{25 \text{ ft}}$$
$$= 2.084 \text{ yd}^3/\text{ft} \quad (2.1 \text{ yd}^3/\text{ft})$$

The answer is (C).

Why Other Options Are Wrong

(A) This incorrect option divided the volume of the blasted rock by the total length of four holes instead of just one.

(B) This incorrect option calculated the area of the rock blasted as a circle with an 8 ft diameter instead of as a square with 8 ft sides.

(D) This incorrect option calculated the depth of rock blasted as 25 ft instead of as 22 ft.

95. The distance from the center of the face unit in question to the top of the retaining wall is

$$h = n_{\text{units}} h_{\text{per unit}} = (3.5)(3 \text{ ft})$$
$$= 10.5 \text{ ft}$$

The vertical effective stress at the center of the face unit is

$$\sigma'_v = \gamma h = \left(120 \ \frac{\text{lbf}}{\text{ft}^3}\right)(10.5 \text{ ft})$$
$$= 1260 \text{ lbf/ft}^2$$

The active lateral earth pressure coefficient is

$$k_a = \tan^2\left(45° - \frac{\phi}{2}\right) = \tan^2\left(45° - \frac{30°}{2}\right)$$
$$= 0.333$$

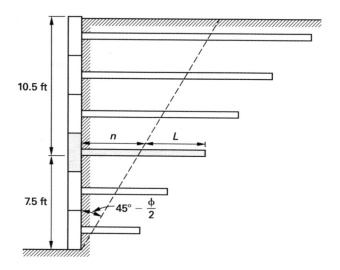

The horizontal stress at the center of the face unit is

$$\sigma'_h = k_a \sigma'_v = (0.333)\left(1260 \ \frac{\text{lbf}}{\text{ft}^2}\right)$$
$$= 420 \text{ lbf/ft}^2$$

The area of the face unit in question is

$$A_{\text{face}} = (3 \text{ ft})(3 \text{ ft})$$
$$= 9 \text{ ft}^2$$

The force on the face unit is

$$R = \sigma'_h A_{\text{face}} = \left(420 \ \frac{\text{lbf}}{\text{ft}^2}\right)(9 \text{ ft}^2)$$
$$= 3780 \text{ lbf}$$

If the metal strip extends to a length of L beyond the active failure plane, as shown, then the surface area of the strip beyond the active failure plane is $(2 \text{ ft})L$, and the frictional force that the metal strip generates is

$$F_f = 2\sigma'_v \tan\phi A_{\text{strip}} = (2)\left(1260 \ \frac{\text{lbf}}{\text{ft}^2}\right)\tan 30°(2 \text{ ft})L$$
$$= (2910 \text{ lbf/ft})L_{\text{ft}}$$

The frictional force must be greater than the active horizontal pressure by a factor of safety of 3. Therefore, the length of strip (in feet) can be obtained.

$$\frac{F_f L}{3} = R$$
$$\frac{\left(2910 \ \frac{\text{lbf}}{\text{ft}}\right)L_{\text{ft}}}{3} = 3780 \text{ lbf}$$
$$L = 3.90 \text{ ft}$$

The length of that portion of the strip within the active failure plane, n, can be calculated using geometry.

$$n = (7.5 \text{ ft})\tan\left(45° - \frac{\phi}{2}\right) = (7.5 \text{ ft})\tan\left(45° - \frac{30°}{2}\right)$$
$$= 4.33 \text{ ft}$$

The total length of the metal strip is

$$L + n = 3.90 \text{ ft} + 4.33 \text{ ft} = 8.23 \text{ ft} \quad (8.2 \text{ ft})$$

The answer is (C).

Why Other Options Are Wrong

(A) This incorrect option calculated the length of only that part of the strip beyond the active failure plane.

(B) This incorrect option calculated the length of only that part of the strip within the active failure plane.

(D) This incorrect option assumed only one side of the metal strip contributes to the frictional force generated by the strip.

96. The cross-sectional area of one concrete pile is

$$A = s^2 = (8 \text{ in})^2$$
$$= 64 \text{ in}^2$$

The volume of one pile is

$$V = AL = \left(\frac{64 \text{ in}^2}{\left(12 \, \frac{\text{in}}{\text{ft}} \right)^2} \right) (50 \text{ ft})$$
$$= 22.22 \text{ ft}^3$$

The weight of one pile, including the driving appurtenances, is

$$W_p = V\gamma + W_{\text{driving}} = (22.22 \text{ ft}^3)\left(150 \, \frac{\text{lbf}}{\text{ft}^3} \right) + 1000 \text{ lbf}$$
$$= 4333 \text{ lbf}$$

The weight of one pile per foot is needed to find the K-value.

$$\frac{W_p}{L} = \frac{4333 \text{ lbf}}{50 \text{ ft}}$$
$$= 86.66 \text{ lbf/ft}$$

From the table given in the problem, the K-value is 0.4.

Find the safe driving load using the equation provided. (This equation is not dimensionally consistent.)

$$R = \left(\frac{2E}{S + 0.1} \right) \left(\frac{W_r + KW_p}{W_r + W_p} \right)$$
$$= \left(\frac{(2)\left(15,000 \, \frac{\text{lbf}}{\text{ft}} \right)}{0.25 \text{ in} + 0.1 \text{ in}} \right) \left(\frac{3500 \text{ lbf} + (0.4)(4333 \text{ lbf})}{3500 \text{ lbf} + 4333 \text{ lbf}} \right)$$
$$= 57,265 \text{ lbf} \quad (57 \text{ kips})$$

The answer is (A).

Why Other Options Are Wrong

(B) This incorrect option neglected to take into account the weight of the driving appurtenances.

(C) This incorrect option used an incorrect value of 0.6 for the K-value.

(D) This incorrect option neglected to take into account the weight of the driving appurtenances and used an incorrect value of 0.6 for the K-value.

97. ASCE 7 Sec. 7.7 covers snow drifts formed from a higher roof or from wind. *Leeward drifts* result from wind blowing in the opposite direction from where the drift is located. The leeward drift height, h_d, is found using ASCE 7 Fig. 7-9 and the length of the upper roof, l_u.

From ASCE 7 Fig. 7-9, if l_u is less than 25 ft, 25 ft can be used as the length of the upper roof. The width of the water tower is 10 ft, so use 25 ft. The ground snow load is 50 lbf/ft^2. Therefore, from ASCE 7 Fig. 7-9, the drift height is approximately 1.8 ft.

The answer is (B).

Why Other Options Are Wrong

(A) This incorrect option calculated the drift height from the windward direction.

(C) This incorrect option misread ASCE 7 Fig. 7-9 by using the value of ground snow as l_u, and vice versa.

(D) This incorrect option assumed that the full height of the water tower would be filled with snow drift.

98. *AISC Manual* Table 3-23 lists formulas for calculating deflections from various forms of loading. The deflections that result from different loads are cumulative. Load factors are given in ASCE 7 Chap. 2. Although LRFD is used, the deflection is calculated based on unfactored loads. The uniform dead load with self weight is

$$w_{u,D} = 0.8 \, \frac{\text{kips}}{\text{ft}} + 0.026 \, \frac{\text{kips}}{\text{ft}}$$
$$= 0.826 \text{ kips/ft}$$

The concentrated live load is

$$w_{u,L} = 5 \text{ kips}$$

AISC Manual Table 3-23 gives equations for shears, moments, and deflections. Since this is a simple steel beam, use case 1 to find the maximum deflection from the dead load. The modulus of elasticity for steel is $E = 29,000$ kips/in^2. From *AISC Manual* Table 1-1, the moment of inertia around the x-axis for a W12 × 26 is $I = 204$ in^4. Using the dead load for w, the maximum deflection from the dead load is

$$\Delta_{\text{max},D} = \frac{5 w_{u,D} l^4}{384EI} = \frac{(5)\left(0.826 \, \frac{\text{kips}}{\text{ft}} \right)(12 \text{ ft})^4 \left(12 \, \frac{\text{in}}{\text{ft}} \right)^3}{(384)\left(29,000 \, \frac{\text{kips}}{\text{in}^2} \right)(204 \text{ in}^4)}$$
$$= 0.065 \text{ in}$$

AISC Manual Table 3-23, case 7, gives the maximum deflection from a concentrated load at the center of a simple beam. P represents the live load.

$$\Delta_{\text{max},L} = \frac{Pl^3}{48EI} = \frac{(5\text{ kips})(12\text{ ft})^3 \left(12\ \frac{\text{in}}{\text{ft}}\right)^3}{(48)\left(29{,}000\ \frac{\text{kips}}{\text{in}^2}\right)(204\text{ in}^4)}$$
$$= 0.053\text{ in}$$

The total maximum deflection from both uniform and concentrated loads is

$$\Delta_{\text{max,total}} = \Delta_{\text{max},D} + \Delta_{\text{max},L} = 0.065\text{ in} + 0.053\text{ in}$$
$$= 0.118\text{ in}$$

The allowable deflection for this beam is

$$\Delta_{\text{allowable}} = \frac{L}{360} = \frac{(12\text{ ft})\left(12\ \frac{\text{in}}{\text{ft}}\right)}{360}$$
$$= 0.4\text{ in} \quad [> 0.160\text{ in; OK}]$$

The maximum deflection is less than the allowable deflection, so the design is OK.

The answer is (A).

Why Other Options Are Wrong

(B) This incorrect option multiplied the dead and live loads by their LRFD load factors.

(C) This incorrect option neglected to convert the beam length from feet to inches.

(D) This incorrect option multiplied the dead and live loads by the LRFD load factors and also did not convert the beam length from feet to inches.

99. The design moment capacity of a reinforced concrete beam is found from the equation

$$\phi M_n = \phi A_s f_y \left(d - \frac{a}{2}\right)$$

d is the effective depth, and a is the depth of the equivalent rectangular stress block as defined in ACI 318 Sec. 10.2.7.1. The area of reinforcing steel for a no. 6 reinforcing bar is 0.44 in^2, so

$$A_s = (3)(0.44\text{ in}^2) = 1.32\text{ in}^2$$

The effective depth is the distance from the compression edge of the beam to the centroid of the reinforcing steel, which is the height of the beam minus the clear cover and half the bar diameter. The diameter of a no. 6 bar is 0.750 in, so

$$d = 18\text{ in} - 2\text{ in} - \frac{0.750\text{ in}}{2} = 15.63\text{ in}$$

The depth of the stress block, where β_1 is 0.85 when f'_c is 4000 lbf/in^2, is

$$a = \frac{A_s f_y}{\beta_1 f'_c b} = \frac{(1.32\text{ in}^2)\left(60{,}000\ \frac{\text{lbf}}{\text{in}^2}\right)}{(0.85)\left(4000\ \frac{\text{lbf}}{\text{in}^2}\right)(12\text{ in})}$$
$$= 1.94\text{ in}$$

According to ACI 318 Sec. 10.5.1, the minimum reinforcement is

$$A_{s,\text{min}} = \left(\frac{3\sqrt{f'_c}}{f_y}\right) b_w d = \left(\frac{3\sqrt{4000\ \frac{\text{lbf}}{\text{in}^2}}}{60{,}000\ \frac{\text{lbf}}{\text{in}^2}}\right)(12\text{ in})(15.63\text{ in})$$
$$= 0.59\text{ in}^2 \quad [< 1.32\text{ in}^2, \text{OK}]$$

Calculate the tensile steel strain using the equation

$$\epsilon_t = \frac{\epsilon_{\text{cu}}(d - c)}{c}$$

c is the depth of the Whitney stress block and, from ACI 318 Sec. 10.2.7.1, is equal to

$$c = \frac{a}{\beta_1} = \frac{1.94\text{ in}}{0.85} = 2.28\text{ in}$$

ϵ_{cu} is the strain at which concrete crushes and is equal to 0.003. The strain in the steel, then, is

$$\epsilon_t = \frac{\epsilon_{\text{cu}}(d - c)}{c} = \frac{(0.003)(15.63\text{ in} - 2.28\text{ in})}{2.28\text{ in}}$$
$$= 0.018$$

This is greater than 0.005. Therefore, per ACI 318 Sec. 10.3.4, the section is tension-controlled. From ACI 318 Sec. 9.3.2.1, the strength reduction factor, ϕ, is 0.9 for tension-controlled sections.

The nominal moment capacity is

$$\phi M_n = \phi A_s f_y \left(d - \frac{a}{2}\right)$$
$$= \frac{(0.90)(1.32\text{ in}^2)\left(60{,}000\ \frac{\text{lbf}}{\text{in}^2}\right)\left(15.63\text{ in} - \frac{1.94\text{ in}}{2}\right)}{\left(12\ \frac{\text{in}}{\text{ft}}\right)\left(1000\ \frac{\text{lbf}}{\text{kips}}\right)}$$
$$= 87.08\text{ ft-kips} \quad (87\text{ ft-kips})$$

The answer is (A).

Why Other Options Are Wrong

(B) This incorrect option omitted the strength reduction factor, ϕ.

(C) This incorrect option used the full depth of the beam in place of the effective depth.

(D) This incorrect option omitted the strength reduction factor, ϕ, and used the full depth of the beam in place of the effective depth.

100. Chapter 6 of the *Manual on Uniform Traffic Control Devices* (MUTCD) gives provisions for temporary traffic control elements. The formula for the merging taper length, L, is given in MUTCD Table 6C-4, where W is the width of the offset, and S is the posted speed limit. (The equation is not dimensionally consistent.)

$$L_{ft} = \frac{W_{ft} S^2_{mph}}{60} = \frac{(12\ ft)\left(40\ \frac{mi}{hr}\right)^2}{60}$$
$$= 320\ ft$$

A merging taper length of 320 ft is needed.

The answer is (C).

Why Other Options Are Wrong

(A) This incorrect option calculated the shoulder taper length instead of the merging taper length.

(B) This incorrect option calculated the shifting taper length instead of the merging taper length.

(D) This incorrect option was obtained using the formula for merging taper length with speed limits greater than 45 mph.